TOMORROW IS TOO LATE

An International Youth Manifesto for Climate Justice

EDITED BY

GRACE MADDRELL

THE
INDIGO
PRESS

THE INDIGO PRESS
50 Albemarle Street
London W1S 4BD
www.theindigopress.com

The Indigo Press Publishing Limited Reg. No. 10995574

Registered Office: Wellesley House, Duke of Wellington Avenue
Royal Arsenal, London SE18 6SS

This edition first published in Great Britain in 2021 by The Indigo Press

A CIP catalogue record for this book is available from the British Library

Image credits: Abner C © Dannine Consoli, Arshak Makichyan © FFF Russia, Asam Sbaih © Sliman Mukarker, Deniz Çevikus © Handan Çevikus, Emmanuel Niyoyabikoze © Yves Manirambona, Fernanda Barros © Luma Santiago, Lucila Huilen Auzza © Carmen Armeya, Maddy Anna Burton © Martha Noakes, Meg Watts © Paula Averkamp, Michael Bäcklund © Tal Rachamim, Nasratullah Elham © Teerakit Vichitanankul, Natalia Blichowska © Barbara Blichowska, Raymond Simpie Smith © Akhona Manyoni and Lubanzi Mazibuko, Rita Naumenko © Zhenya Zhulanova, Saoirse Exton © Séan Curtin, True Media, Vanessa Nakate © Keith Kaliisa, Flynn © Rebecca Naen/Guardian News & Media.

ISBN: 978-1-911648-32-1
eBook ISBN: 978-1-911648-33-8

Design by houseofthought.io
Typeset in Goudy Old Style by Tetragon, London
Printed and bound in Great Britain by TJ Books Ltd, Padstow

This book is dedicated to my best friends,
to Kyle, Lucy, Patsy, Dan, Anna, Nat,
Maude and Helen, for being the best people on
the planet and the brightest stars in all the sky.

—G.M.

CONTENTS

GRACE MADDRELL first went on school strike for the climate at the age of thirteen, and has since become a passionate activist for equality and climate justice. Grace was first inspired to tweet asking for contributions to what became *Tomorrow Is Too Late* in November 2019, two months after marching in the September global climate justice strike. Grace has worked with various activist groups, including Fridays For Future and #SaveCongoRainforest, and helped co-found Solo But Not Alone, a social media solo striker support group. Grace is homeschooled and lived in Somerset in the southwest of England at the time of compiling *Tomorrow Is Too Late*. Find Grace on Twitter: @ElmGrace.

INTRODUCTION

Firstly, I want to take a moment to thank you for picking up this book. Thank you for seeing that it contains voices which need to be heard. I can't believe you are holding this in your hands! It has been such a ride to get to this point, both in my life and specifically with *Tomorrow Is Too Late*.

My name is Grace, I'm fifteen at the time of writing (December 2020), and I'm a climate and equality activist from England. I work with the #SaveCongoRainforest campaign and Fridays For Future, as well as speaking up independently. I am one of the co-founders of Solo But Not Alone, a social media group dedicated to supporting and promoting climate activists who have to protest alone. I am also the founder of Queer and Here For You, a support and advocacy group for my fellow LGBTQ+ youth, and I speak up online for all forms of equality and justice.

To some extent, I have been aware of the climate crisis for ages. When I was little, my dad would donate to Greenpeace and so on, and I knew that this was bad, this was dangerous. But I think the difference between knowing on an abstract level, and waking up to the genuine danger that many are facing, that we will all face if this crisis isn't averted, is very different.

When I was thirteen, in February 2019, some classmates were discussing a climate strike. One of them asked me if I wanted to come along. I said I would. After I realized this protest was going on, I started to think a lot more about the climate crisis. By the time the strike came along, in March 2019, this emergency was at the front and centre of my mind, a lot of the time. For the entire month that elapsed between hearing about the strike and it actually happening, I had been thinking about it, preparing for it, and watching speeches by Greta Thunberg, the founder of the Fridays For Future movement. She started activism in August 2018, and kicked off youth strikes around the world. Although the media tries to focus solely on Greta, there are millions of young people all over the world who are standing up for the climate. On 20 and 27 September 2019, around seven *million* people marched for the climate, all over the world. Some of those who marched were veterans in this fight, activists who had long been part of Greenpeace, or Indigenous elders who had been advocating for decades. Others were first-timers, only just realizing the power their voice could have. But no matter who, where, or how many times they'd protested before, everyone who striked on that day helped make it what it was – one of the largest mass mobilizations ever. And although it was a general strike, attended by people of all ages, youth were the ones leading the calls for change.

But back to March 2019, and my first nervous foray into activism. That first march that I attended felt like I had finally found where I belonged – in a group of passionate young people shouting for change. I wish every moment in activism could be that empowering. Shouting 'What do we want?' 'CLIMATE JUSTICE!' as we marched along the street, waving banners and yelling ourselves hoarse, felt like… it felt like I was actually doing something for once, rather than just stumbling through my life. It felt like maybe there was a point to my existence after all. And it felt like I was finally able to let out my anger at all the corruption, at the world leaders and at corporations, at our corrupt, capitalist, unequal society.

Since then, it has been a huge journey to get to where I am now. I don't mean in terms of fame or even campaigning, but to where I am in my life. I started activism as a closeted, doubt-filled thirteen-year-old with no real friends who merely dreamed of being an author and was desperate for my name to be known. Now I am soon to be published, friends with the most incredible people in the world, out and proud as LGBTQ+, and I know that there are far more important things than being known, not to mention that there are so many others on the front lines of this crisis who need to be heard more than I do.

I struggle still, of course. Activism is hard. I've been doing a lot less during the Covid-19 pandemic, but even organizing calls and book editing can be tiring, and mental health can suddenly go out of the window for no reason at all. But what keeps me going is knowing about all the wonderful people fighting with me – and so many of them are incredible.

This book was inspired by *Resisters*, a book by Lauren Sharkey, which collected the voices of young female activists from around the world. After reading that, I thought that I would love to put together one specifically focused on climate activists. And I decided to open it up to youth of all genders, too.

I started this book simply with a Twitter post, in November 2019. Within hours, messages were flooding in from interested people. I also reached out to a lot of young activists myself. Of course, not everyone who I messaged or who expressed interest ended up in this book. Many people never got round to writing an essay at all. Those you read here are those who did. But the whole process was really amazing. Around five months after I first started work on the book, The Indigo Press accepted it, and I realized that this was really, really happening. Working on it hasn't always been easy. The hardest part has been having to pressurize people to get in their essays, edits, etc., to meet a deadline. Pressure is definitely among my least favourite things! The loveliest experiences that I gained through this were mostly about bringing together these young people from across the world, hearing their stories and journeys and being able to (hopefully) amplify their voices. Some of the essays are really sad, of course, but they're all so beautiful.

I've learned a lot through this. My essayists have taught me so many things about how the climate crisis affects them, their country, their demographic. The actual publishing process has also been extremely educational for me. All of it has been something that not many people of my age would get to experience, and I am so, so grateful that I have.

Some people might wonder who I'm hoping will take notice of this, who I want to see it, and why. But my answer is: you. Whoever is reading this. All of you. I want *you* to see it. I want everyone to be able to read the words of these activists, because their stories are so important.

Before you read their words, though, there are a few practical things to say about this book.

Firstly, the ages included here are the ages at which they wrote their essays. By the time you read this, someone who was seventeen when they wrote their essay might be twenty-two. That's why their years of birth are also included in their bios to give you an idea of what sort of era they're writing in and how old they are now.

Secondly, although all these essays stay absolutely true to the fundamental message that their authors are trying to get across, they have all received varying levels of editing to make sure that they make sense and are as good as they can be.

Thirdly, the essays for this book were compiled across an eleven-month period. Thus, the perspectives are all different. Some people will discuss the Covid-19 pandemic, while others might talk about 'recent' Australian bush fires, for example.

Fourthly, all names have been included in the format that the essayists preferred. For example, some didn't want their surname to be included, and others only wanted their last initial rather than a full name. Individual wishes have, of course, been respected as regards all names, pronouns and amounts of information included.

Fifthly, do bear in mind that many of these activists are writing in their second, third or fourth language. So huge props to them for using English for their essays. English is the dominant language of climate activism, which is another form of discrimination against many activists – several of the writers had to seek out information in English because there was so little available in their native language.

Sixthly, all photos included are done so with permission and relevant credits, and where a photo is not included, it is due to the contributor not wishing to have one in the book.

Finally, I would like to make it clear that most of my proceeds from this book will be going either to help those who need it or to organizations working for the climate and/or for equality.

I think that's it for the technicalities.

I compiled this book because I know that youth voices need to be heard. Leaders often try to ignore young people, and to silence us. In the countries on the front lines of the climate crisis, young activists are routinely silenced by either their governments or by the media, which prefers to focus on activists from places like Europe and North America. And for those of us who are lucky enough to not yet be facing the serious effects of this crisis, we will be the ones who are around to experience them.

So, I hope that you will read the messages of these activists, listen to them, and take action. I hope that this anthology has succeeded in bringing together youth from all around the world, and in making sure that their messages can no longer be ignored.

Because we young people are powerful. We deserve to be heard and listened to. We deserve a say in our future. The leaders – you, you leaders – you are selling what was never yours. The lives of millions. Including us, your children. But we are not going to sit by and let this happen. We are acting. I hope this book will inspire everyone to do the same. Across the globe, amazing youth are standing up for their right to a planet. And I am so proud to be working with these people to create change. I am so honoured to be able to share their voices and stories here.

The climate crisis is a huge, overwhelming issue that has implications for every aspect of our lives. In *Tomorrow Is Too Late*, youth activists talk about many of these issues, relating the impacts the crisis has already had on their mental health, local communities, food security and childhoods, to mention just a few.

In tackling the climate crisis, we also have to tackle the underlying factors of colonialism, racism, capitalism and inequality that created the issue in the first place. And, of course, including voices for/information from any of the countries and areas in this book does not in any way equal support for their regimes, political status, or current state of existence.

Long before 'climate change' was even something people knew anything about, many were suffering every day due to the same things that caused this crisis. We cannot continue in a world like the one we live in today, a toxic world where those who lead are very often those who choose to treat some lives as more important than others. Those living in systems built around prejudice and privilege. Although the climate crisis will exacerbate all these existing problems, it is a symptom of how fundamentally broken our society is, rather than the cause. Everything needs to change, and it needs to change soon, because time is running out, and it's running out fast.

A better life, a brighter future, an equal world, is possible, but we have to work for it. And that's what all the young people in this book are doing. From Palestine to Kenya, from Argentina to Australia, and from Germany to Canada, these activists are changing things, one word, one step, one march, one call, one article, one lesson at a time. No matter what form their activism takes – from founding movements to picking up plastic to simply advocating online – the youth of *Tomorrow Is Too Late* are making the world a better place. I hope their stories show you that you can, too. Because every single one of us can make a difference. And you don't need to wait until you're an adult, or until there's an established movement in your area, or until your friends do it first. Even a small action that you take is better than no action. You may think that you're just a drop in the ocean. But without the drops, there is no ocean. So use your voice and become part of the wave of change right now, because tomorrow is too late.

GRACE MADDRELL
TWITTER: @ElmGrace

ABNER C, 11

USA

Abner (born 2008) is a climate striker from Oregon in the USA, who's been striking weekly since July 2019. As well as protesting, he writes letters about the climate crisis to every state governor across his country. Since December 2018, he has written over a hundred letters to elected officials. Sometimes he gets a reply... sometimes not. He has also been labelled a UN Protector of Progress. As well as advocating for general climate action, Abner also highlights specific issues, like the fires in the Amazon and Congo rainforests.

Portland, Abner's city, is one that has a reputation for being 'sustainable', but, as of 2020, it has not even declared a climate emergency, unlike many towns, cities and countries across the world. And, in the eyes of this inspiring young activist, Portland, like nearly everywhere else, 'has not done enough to fight climate change'.

Luckily, Abner's family are very supportive of his activism, and take him to the city hall on Fridays so that he can fight for Portland to do better. Additionally, Abner has a younger sister, who joins his strikes about twice a month.

On 20 September 2019, Portland held the third-largest climate strike in the USA, and Abner was one of its organizers. Undeterred by the crowds, he spoke, bravely and strongly, into the microphone on the stage. 'We know what we need to do, and we just need to do it,' he said. 'If everyone understood what is happening, we could have overcome the climate crisis by now... Sometimes we have to start something, even if we don't know every little detail. We are in one of those situations. What can you do today to help fight the climate crisis? It might not be much, but imagine if we all did that little thing. Then, it would be a big thing.'

Because little things add up to big ones. And solo strikers add up to a movement. And just as no action is too little, so no person is too small, or too young. Which is why, even though he was only eleven years old when he wrote his essay, Abner isn't afraid to stand up, speak out, and make a difference.

TWITTER: @Abner4action

My First Year as a Climate Activist

I became a climate activist on 12 December 2018, days after turning ten. Of course, I didn't know I was an activist at that time. It started with letters about climate change and renewable energy to Governor Kate Brown of Oregon, and Director Benner at the Oregon Department of Energy. I wrote forty-nine more letters before my first climate strike, in front of the Oregon State Capitol, on 29 March 2019. I didn't start striking weekly until July 2019.

Later, I helped organize the local 20 September 2019 global strike with youth and environmental organizations. I worked mostly on the logistics for the festival. On the day of the festival, I gave a speech talking about my climate activism. I spoke about why climate action matters to me. 'Climate action is important to me because the big effects of climate change will hit my generation and the one after that,' I said. I could expand on that a lot, though. There are many other reasons why climate action matters to me. Like the fact that due to climate change, we are ramping up to a sixth mass extinction, or that the fate of humankind simply depends on our action, and much more.

I strike (nearly) every week, and it feels like I am barely noticed. No city official has ever come out and talked to me. They have engaged a bit with some other youths in the area. Some people that walk by give support, others yell and curse. We need drastic change to stop this crisis. We are not doing that right now, it is as simple as that. We are just not trying hard enough. My city needs to lead, so others can follow with specific actions. I joined Fridays For Future, a non-partisan grassroots organization dedicated to stopping climate change. Then I started a local chapter. During my striking, I also try to highlight problems that I learn about from climate scientists or other activists, like the Amazon or Congo rainforest fires.

I have written more than one hundred letters regarding climate change topics, including one to the governor of every state and territory within the United States. I plan to write to them all again, to see if they have made any progress this last year. Mainly I asked them to keep to the commitments of the Paris Climate Agreement, but I have also asked governors, mayors, presidential candidates and other government workers/officials to increase renewable energy, protect the outdoor world – like our national parks – protect wildlife – like butterflies – and do anything they can to curb CO_2 emissions. I also asked presidential candidates to hold debates focused on climate change, before the recent election in the United States.

When I look at what the closest city to me is doing, it is not moving fast enough. Portland is looked at as an environmental leader by other cities in the world. Of course, they have done many things compared to some other cities, like having a youth council, but it isn't enough. Portland hasn't even declared a climate emergency. They must do better! They could stop oil trains, help local transportation move to electric buses faster, support more public transportation and no youth fares, and change some of the 2015 Climate Action Plan targets to 2030, instead of 2050. But they have not done any of those things.

Climate change is the biggest issue of our time, probably the biggest in history. Yet companies pay lobbyists so that our government does nothing. The large oil corporations know about climate change, and its relation to burning fossil fuels, but they still sell it to make money. Climate change has become a partisan issue in the United States. We must act fast, extremely fast.

You can climate strike too. It may feel like you are not noticed, and your action means nothing, but every action will help. Everyone must do their part. If you start acting, you will inspire others around you to take action. The climate crisis is a crisis everyone must be worried about. It might not affect you now, but it will later.

Here are some suggestions if you are, or want to be, a climate activist:

- When kids write letters, it is powerful, especially with drawings.
- Chants are also powerful, record them for social media.
- Be positive when you are striking, more people will listen.
- Take breaks if you need them, post a quick photo for a digital strike instead.
- Be ready to inform people on how they can specifically help, for example, show up to a meeting about fareless transit.
- Remember no one likes to be yelled at.
- Remember to stay informed by reading trustworthy articles.

So, in conclusion, I have been a climate activist since late 2018, and very little has happened since then. It is vital that we do more. Big companies and governments knew about climate change, but chose to ignore it. Your action matters, so please do what you can to fight climate change. You acting to fight climate change does not necessarily mean leaving school or work every Friday and holding a sign. Even small actions matter. So please join the fight for the climate.

ALI KHADEMOLHOSSEINI, 18

Iran / Germany

Growing up in Iran, Ali Khademolhosseini (born 2001), or Amirarian, as he is commonly known, knew that speaking up could put him in danger. However, that didn't stop him: he was labelled a troublemaker at elementary school for standing up for a fellow pupil, and a few years later was expelled from a boarding school under a military administration.

After Amirarian moved to Germany, his activism took a different turn. He joined Fridays For Future, and a year after that co-founded a FFF chapter in his home city. Shortly afterwards, he basically ended up founding a continent-wide group by turning his private Instagram page into Fridays For Future Europe. He also initiated the first FFF Europe meeting, and then it was his idea to start SMILE, a large meeting of European strikers.

As if this weren't enough, Amirarian decided to try to bring the FFF movement to Iran. And he succeeded, at least for a while. The strikes were small, but they were there. However, Iran is a very risky place to be an activist, and several strikers ended up in prison as a result. They were later freed with the help of a lawyer who Ali's uncle paid for.

After this, a new group was started, the Activists in Risk Zones international working group, for which Ali later became the spokesperson. They helped activists all over the world.

Then, disaster struck. After the loss of three of their activists during protests against the Iranian regime, Fridays For Future Iran, grieving and endangered, decided to dissolve itself.

Ali has directly experienced tyrannical regimes and dangerous situations. But still he keeps going with his activism. I hope that his bravery will be remembered. I hope that, as is the case with many in FFF Europe, Ali Khademolhosseini's name will become known. Because it deserves to be.

INSTAGRAM: @khademohali / TWITTER: @KhademOHAli

Perhaps I had one of the strangest motives to join Fridays For Future – having a crush on someone who is an eco-activist! I'm going to spoil the end. Nothing happens between me and her, because I fall in love with Fridays For Future.

I am Ali Khademolhosseini, or just simply Amirarian, and this is my story.

I love physics and because of that I know things about nature, but more than that, my life is built around politics and activism.

When I was four, I was sent to learn judo and jiu-jitsu. I sometimes think that my interest in politics started there. I learned to speak up and defend the defenceless.

When I was ten years old, me and other students had a bad fight with the school council, over a student getting expelled from the school because of criticizing the supreme leader Khamenei. Yes, children in Iran, the country where I was raised until I was fourteen, have a sense for politics.

Anyway, back to the story. So he got expelled, and because he insulted the supreme leader at gym we all got detention. On our way back to school, I raised my voice at the sports teacher because, before the student I am talking about got expelled, he was slapped on his left cheek by him. I told him how humiliating and inhumane his behaviour was, and for that I also got slapped.

Anyhow, on the way into school, we planned a strike, where the ten-year-olds would strike for a week in the schoolyard, in solidarity with our friend who was expelled because of criticizing the supreme leader.

From that moment a sense of being was activated in me, and I saw how leadership works.

After finishing elementary school, I was sent to a military school which belongs to the Iranian national air force. I got my name from there. When I was fourteen we played paintball and our team was called Iran, and there were twenty-three of us, and we won. And since I was leading the team (although it was the teamwork which got us to the victory), I was named Amirarian, which means 'the commander of Iranians'.

Two months after that game I was kicked out of school, because of promoting anti-regime ideologies and being a freethinker, someone who seems to be liberated from religious ideologies and who lives their life on the principles of freedom of thought. And a month after that I was in Germany because my parents were working in Germany all of a sudden. And I am thankful for being in Germany. Who knows what might have happened to me, with the dangerous personality I have, if I were still in Iran?

Until the beginning of 2018, my life was, mainly, boring. I barely could speak good German, and I was kind of socially isolated, but I still had the activist spirit alive in me.

I first heard about Greta Thunberg through Instagram. I noticed right away what the topic was, and although I knew about global warming and climate change, it was not so obvious to me, before I learned about Greta, what was actually happening.

I became active on 3 January 2019. I co-founded Fridays For Future Erlangen with another activist, and three weeks later we had our first strike, with over five hundred people.

All of a sudden, I felt alive again.

I do not believe in the validity of any geopolitical borders, therefore I wanted to have a global impact. On 23 January, I was already active at an international level. Shortly after, I turned my private Instagram page into Fridays For Future Europe and with that, I started the idea of working on a European level together with some famous activists, in order to develop our structures faster and make our impact on politics and our societies much deeper.

At the same time, I was so in love with Fridays For Future that I did everything to bring the movement to Iran. With all the difficulties we have in Iran, we had our first strike, with five people, in Tehran, and two weeks later we had over four local groups in Iran. Everything was going surprisingly well for a while.

My love of the movement got even stronger after I organized and coordinated the first European meeting in Strasbourg. The Green Party invited us to the European Parliament, and there I met many friends who I just knew from the Discord chat groups or from one of the hundred groups on the different platforms we have. People like David Wicker, Isabelle Axelsson, Vivian from England, many others who are now famous.

The meeting in the European Parliament was a success, since I, at least, produced the idea of SMIL – which stands for 'Summer Meeting In Lausanne'. However, about a month after 13 March 2019 SMIL turned to SMILE. The E stands for Europe – this idea was from Loukina, a climate activist from Switzerland.

In the same period of time something bad happened in Iran. Twenty strikers disappeared after they were arrested in Tabriz, a city in the northwest of Iran. Four days later, a friend of my uncle called me. By the way, I was using all my resources to find out what happened with them.

Back to the story…

They were in Evin prison in Tehran, and my uncle paid for a lawyer, who got them out, with many difficulties.

This situation was the reason for the birth of the Activists in Risk Zones international working groups (ARZ IWG), the purposes of which are protecting human rights, helping activists in dangerous situations, and above all, providing security and safety education and advice for them.

I was later elected as the spokesperson of the ARZIWG, but this was actually the beginning of the group. Very soon after this, we reached out to all the national groups in the risk areas and told them how to document and how to calculate the potential threat.

We helped India to get the word around, we observed a sixteen-year-old activist from China, because she felt threatened, and also helped imprisoned activists in Kenya to get out of prison.

Being an activist is not easy. If you are someone who lives in Iran, you have to always look back over your shoulder, to check you are safe.

In November 2019 we faced the saddest event in the history of Fridays For Future, although many do not know that it actually happened. In that month, thousands went on the streets of Iran to protest against the tyrannical regime of Iran. But the regime opened fire at the protesters with live ammunition. More than 1,500 people were killed, and 70,000 were arrested.

You might ask what does this have to do with Fridays For Future? The answer is short but important. Sassan, Muhammad and Maryam, three Fridays For Future activists in Iran, were killed in those protests. We Fridays For Future activists see our relationships with each other as the relationships of an extremely big family.

It is always sad to lose your sibling. Losing them was, for us all, depressing. I believe that the war for climate justice, without fighting authoritarian, tyrannical and undemocratic forms of state or government, is meaningless, and we will actually gain no success without winning that battle as well.

After their deaths, the activism became truly Mission Impossible. Not only did their deaths make it difficult, they also raised a major security risk for all our activists. Our network was jeopardized by the Basiji forces, a part of the Islamic Revolutionary Guard Corps, which is a regime known as a terrorism agent in Iran and outside Iran.

After Fridays For Future Iran dissolved itself, I lost the feeling of belonging.

I did not know and still, as I write on 29 March 2020, do not know, what national group I belong to. Because of that I always identify myself as a part of Fridays For Future International. Fridays For Future has been the most important event of my life. I sincerely believe that some of the experiences I have had, in Fridays For Future, are experiences that might have taken years otherwise.

I just encourage everyone to join grassroots movements to actually learn something from life.

ANGIE MASSOUD, 14

Egypt

Angie Massoud (born 2006) is a teenage climate advocate and songwriter from Cairo, Egypt. Because in her country there is, in her words, 'not enough freedom' to join protests and marches, Angie is mainly active online. Since December 2019, she has been raising awareness on Twitter, and has also got involved, online, with Greenpeace International and the Climate Reality Project. She is also working on a song about the climate crisis.

Angie decided to get involved with the climate movement to 'support communities seeking climate justice through the power of law'. She signed up to join Greenpeace on 10 December 2019 and helps with their online campaigns.

In February, when she was to go back to school, Angie planned to do her first protest. She wanted to pin a sign to her bag, telling her classmates about the climate crisis. But sadly, when the time came, Covid-19 had hit Egypt, and Angie had stopped going to school out of fear of the virus.

However, this didn't stop her from continuing to advocate online, amplifying others' voices through retweets as well as posting messages of her own. She also kept designing graphics and videos, as she had been doing since December, to spread simple but powerful words about the climate crisis. For instance, 'The best Valentine's gift [is] your love for your only home. Love your mother,' and 'We did this and we can fix it again. Our planet is our responsibility.'

Despite not being able to strike herself, Angie is always standing in solidarity with those who do. In fact, one of her first tweets told everyone to keep going.

It doesn't matter that Angie's advocacy is nearly entirely online, because in our current society, that is the place you are very likely to be seen. That is the place where many of us know people from. And that means that Angie, despite not being able to march on the streets, is, most definitely, making a difference.

INSTAGRAM: @angiemassoud / TWITTER: @angie_massoud

Egypt is one of the countries most affected by climate change. It has altered our life, our vision for the future and our natural environment too, not just with one impact but with a lot of impacts. I wrote about them on social media, trying to discuss what is happening, what we should do, and how climate change is affecting Egypt, but unfortunately only a tiny number of people seemed to care. I'll talk about some impacts that are caused by climate change in Egypt.

ECOSYSTEM

Climate change has destroyed 50 per cent of the coral reefs in the Red Sea. Some of divers predict that the coral reef will decline by 80 per cent by 2060 and that there will be a rise in sea levels, which will lead to flooding in the lands on the northern coast of Egypt and will lead to a decrease in the bird population.

WATER RESOURCES AND FARMING

Egypt's National Strategy for Adaptation to Climate Change and Disaster Risk Reduction recorded in 2016 and 2017 that by 2100 the sea level will rise about 100 cm. This will result in mixing groundwater with salty water and this will affect agricultural crops' quality and production processes.

EGYPTIAN HEALTH

Egypt is suffering from ever-higher temperatures; in the summer the temperature can rise to 50°C and many people can't handle this, so it causes premature deaths.

High temperatures also cause an increase in, and proliferation of, the insects that carry diseases such as malaria, bilharzia and hepatitis C. And because of these high temperatures people increase their use of air conditioners and fans and this is leading to frequent power cuts. The polluted air that the people breathe is a result of using fossil fuels in industry.

Many Egyptian people have a really bad habit – they throw their trash in the streets and then a volunteer comes, collecting the trash and then burning it. They think they're helping but they don't know that this is helping to pollute the air, and it has a negative impact on their lungs.

With all these negative impacts, climate change will affect Egypt's economy too. It is likely to lead to losses of 20 billion Egyptian pounds by 2030, and this is so soon, and will affect all the Egyptian people, and these losses will increase by 2060 to 122 billion.

Egypt is dealing with climate change and establishing projects that help to protect the country from a lot of impacts, and here are two of them:

- Egypt is working on using clean energy in transportation instead of using fossil fuels, and developing projects to protect the Egyptian Delta.
- Egypt encourages energy efficiency improvement projects to encourage investing in clean energy, including waste treatment and establishing new arboreal forest.

I hope we can protect our world and fix these problems without any dangers for the next generations.

Our planet needs every person on it to work to help our only home for a better life, health and future.

Don't wait for other people to make a difference. You can make a difference no matter how little you are. You may have a small idea but you don't know how it will help your country and change minds.

Youth started acting on the climate crisis, and now is the time for our leaders to start.

ARSHAK MAKICHYAN, 25

Russia

Arshak (born 1994) is an activist from Russia. He was born in Armenia and moved to Russia at the age of one. He is a violinist, and a vegan.

He started striking in March 2019, and for a long time he was the only Russian striker. But then he banded together with some other activists and started to create a movement. Now there are several cities where strikes occur every week.

His country is one where it is hard to protest, and the climate crisis is quite a 'new topic' there, because most of the mass media were silent about it for years and are only now starting to wake up. Arshak himself has, at the time of writing this essay, been detained twice. The first time was when he joined protests about the Moscow election on 27 July 2019, and the second was on 25 October, when the government refused the groups a mass strike but they decided to strike anyway.

Arshak was imprisoned for six days. In a YouTube video, he said, 'Jail in Russia is quite a complicated thing, with torture and everything, but there are thousands of people dying and suffering because of the climate crisis.' He showed incredible bravery, saying, 'It's not so important what will be [happening] with me,' as compared to the severity of the climate crisis.

Very few people of twenty-five would have the courage required to face the potential of arrest, imprisonment and even torture, for the planet's sake. But Arshak did. So his story needs to be heard. Peaceful protest is a human right, but in Russia it is not treated that way. To help, use the hashtag #LetRussiaStrikeForClimate. And if you are privileged enough to be able to protest peacefully without the fear of being arrested just for standing there with a sign, then always remember that some people don't have that privilege, and yet they still choose to stand up and speak out. If a twenty-five-year-old can do it, facing the potential of arrest and torture, then the rich and privileged older people who created this crisis certainly can.

Courage doesn't even begin to cover it. Arshak is truly a climate hero.

INSTAGRAM: @makichyan.arshak / TWITTER: @MakichyanA

I was striking alone for a lot of weeks in Russia, but it's not because I was the only person concerned about the climate crisis. There are a lot of international media companies which are writing that I'm the only Russian striker, but it's not true. Not any more. It *was* true for many weeks, but now there are other brave people joining our movement.

But why are we so few?

Almost all our media were silent about the climate crisis at the beginning of 2019, for several different reasons. And even if you were an eco-activist you would be afraid to strike for the climate, because your friends wouldn't understand you. We started the climate strikes from nothing, while people in other countries had a foundation. Nothing about climate change is taught in our schools, either.

And, of course, you have to strike alone in Russia, because it's the only way to protest there without approval from the government.

A little about the history of Fridays For Future (FFF) in Russia.

When I was organizing the second global strike with the other activists, in May 2019, I launched an internet campaign, #LetRussiaStrikeForClimate, as some people in Russia are afraid to protest, and they need some support. And I think it can help us to unite, and to understand that we all live on the same planet, and we have to support each other.

On the second global strike, there were strikes in ten cities in Russia, but the local government refused us permission for a mass strike in Moscow... the richest city in Russia.

After that, we translated the Fridays For Future international manifesto and we decided to organize FFF Russia like it has to be – a grassroots movement. There are no leaders or vertical structures. It's quite an unusual thing for Russia, and it's very important for building an efficient campaign and a new society. Because FFF is not only about protests, we are also something bigger.

Then other cities started to join our weekly strikes. In some cities, it's easier to get authorization. For example, there are weekly mass strikes in Kaliningrad. There was a girl from Gus-Hrustalny, she was striking every week as well. People even started weekly strikes in Archangelsk, the region where the government was going to send millions of tons of garbage from Moscow.

There is FFF in Vladivostok, which is seven days on the train from Moscow. And other cities, where people are striking despite the local governments trying to intimidate them. On 27 September 2019 there were strikes in about thirty cities by 700 people.

There are a lot of cities in Russia where you can see the circumstances of the climate and ecological crisis. And most of the people don't know that they can do something. So we are trying to unite everyone in Russia and describe to the people what is wrong with the climate and what we can do. Initially, there was almost no information about the climate. Now some independent media are writing about it... but why? It was because we were detained, after the Moscow government refused us a mass strike. We have a community that is concerned about the climate crisis. It's small, but they are really great.

Every striker in Russia has a lot of problems. And besides, living in Russia is quite a difficult thing. Especially in smaller cities. But there is hope in Russia. Though it's easy to get desperate living there.

Being an activist is not about doing miracles. It's about small steps. And we are doing a lot of small steps.

A lot of people ask me how they can help me. I don't know. But, of course, you can stop buying fossil fuels, you can help other countries which are much poorer than Russia. And I'm sure that you are going to do that. Because millions of climate activists are watching you. And helping you.

ASAM SBAIH, 24

Palestine

When she was at university, Asam Sbaih (born 1995) was part of the Ecology Friends Association, an environmental group, through which she was able to take part in campaigns for the climate. However, after leaving uni, she found that there was very little opportunity to engage with environmental issues in Palestine, as it lacked climate groups and many of its residents were struggling so much daily, under the Israeli occupation, that organizing anything of that kind was difficult. As a result, Asam is unable to be very active around climate issues, but she still retains a deep love of, and respect for, the planet.

Just because Palestine is a place where it's hard to fight for the climate, it doesn't mean Asam doesn't know as much as the others in this book about the issues facing the world. The people of her country are on the front lines of the effects of the climate crisis, suffering not only from disasters such as droughts, but from deep environmental injustices brought about by the occupation of their land.

Despite all the difficulties faced each day by the youth of Palestine, many of them care deeply about the crisis affecting the whole planet. There is even a small strike group, called Sindyan, in one part of Asam's country.

And the climate isn't all Asam has spoken up for. She previously worked as a coach around gender issues, another area where there is not only inequality and injustice but which is also connected to the climate, as people who are marginalized and discriminated against due to their gender, or other factors such as race, religion or sexuality, will suffer, and are suffering, the most from the climate crisis.

The current situation in Palestine is pretty dire, and it will only get worse if action is not taken on the climate and ecological emergency. However, the country is consistently ignored as regards climate issues, due to its current circumstances (the Israeli occupation).

Asam's essay may be brief, but in it she manages to convey, in just a few paragraphs, the urgency of this crisis – for Palestine especially, but also for the world.

INSTAGRAM: @asam_sbaih / TWITTER: @AsamSbaih

My name is Asam Sbaih, and I am Palestinian. I have a major in biotechnology from An-Najah National University. I coach women in the distant villages in the fields of gender equality and women's reproductive and sexual rights.

My academic background has always helped me regarding ecology and biodiversity, since I participated in ecology clubs like the Ecology Friends Association, for which I worked on several environmental issues including biodiversity and waste recycling and reusing. Being active in such voluntary work has increased my passion towards the environment as well as my interest in learning more about its critical issues.

Throughout my participation in this club, I was involved in different activities and campaigns focusing on biodiversity and environmental protection with the help of academics, specialists and other activists. This in return brought to my attention serious environmental issues like the climate crisis. I became more aware of the life-threatening risks on our planet as a result of misusing natural resources, besides violating laws of nature.

Added to this, I became more aware of the serious Palestinian environmental situation. Especially that Palestinians struggle with severe water scarcity, heavy droughts and other worsening living conditions. Palestinians also face several stark environmental violations caused by the Israeli occupation, such as controlling natural resources, and unfair water distribution.

One of the activities that strengthened the contribution of youth in the climate issue was the participation of a Palestinian environmental youth group called Sindyan in the global campaign Fridays For Future, who held a strike in cooperation with school students, in Ramallah city. The event was not as huge as it may sound, but for sure it was a starting point in the journey of taking action and raising awareness.

Palestinians depend only on the international community to act since we live in a small country that faces dozens of serious existential problems. Our political situation does not allow us to shape international agreements, yet we will be on the front lines when the climate crisis unleashes all its devastating effects. In fact, we are experiencing many right now.

In the end, I really believe that our life and the lives of the next generations are our own responsibility now, and we need to continue to take action to make a real change.

AYISHA SIDDIQA, 20

Pakistan / USA

Ayisha (born 1999) is a co-coordinator of Extinction Rebellion Universities US, and she helped to organize the New York City climate strike on 20 September 2019. She is also one of the co-founders of Polluters Out, an international youth climate justice coalition led by Black people, Indigenous people and other people of colour, which aims to kick polluters out of spheres of influence. For instance, one of their aims is to stop fossil fuel companies from sponsoring the COP climate summits.

Ayisha is Pakistani American, and she moved to the USA at the age of six. Once there, she was exposed to a wide range of discriminatory behaviours and she felt that those in power were trying to silence her voice and erase her identity and religion.

Pakistan is very vulnerable to the climate crisis, even though the country contributes little to the greenhouse gases (GHGs) that cause its suffering. This is one reason for Ayisha's activism.

Another motive that drives Ayisha is that she is a resident of Coney Island, a community in New York City which has suffered some of the most detrimental effects of climate breakdown, and she has seen first-hand how little aid her area has received to repair the damage caused.

These events have led her to fight for climate justice, which means that she doesn't just stand up for ecosystems and the planet, but also for global equality and for the people who have been most affected by this crisis.

Ayisha, and the campaign she helped found, is nothing if not determined. When their planned street-based actions were made impossible by the Covid-19 pandemic, Polluters Out took their message online, raising awareness through Twitter storms and digital strikes, along with various other campaigns.

Institutional prejudices, especially against people of colour and Indigenous communities, are massive contributors to the climate crisis. And without tackling these systems of oppression and discrimination, as the amazing Ayisha and so many other activists are trying to do, what kind of world are we saving?

INSTAGRAM: @ayisha_sid / TWITTER: @Ayishas12

The climate movement in NYC is a collation of youth organizations, in the larger northeast area. I am the co-founder of XR Universities, which is why I was one of the main organizers of the march on 20 September 2019. This demonstration was separate from the weekly strikes done by students affiliated with Fridays For Future and is also separate from the civil disobedience actions taken by XR participants.

My role in organizing the 20 September strike varied from doing outreach (handing out flyers), to meeting with press and UN envoys and writing letters to our city government to give public school students excused absences on 20 September. I was responsible for getting college youth to show up to the march, which they did (in thousands). After the strike, I spoke at multiple panels, including one at the Cooper Union for the Society for Ethical Culture, and was invited to have conferences with Al Gore, Tom Steyer (the 2020 presidential candidate), the Sea Shepherd sea conservation organization and Jane Fonda.

The significance of the climate movement for me has a lot to do with my background. My early childhood was spent in the province of Jhang, by the Chenab River. My upbringing emphasized the symbiotic relationship that Indigenous peoples, and peoples of the land, have with the Earth. Medicines have been passed down for centuries, and an internal intuition allows them to determine how to take care of the soil, and the right amount of fish to take from the river so as not to disturb the ecosystem. They understood the mechanisms of homoeostasis long before the word made its way into our biology textbook. I grew up watching the people around me walk with humility alongside the creatures of the Earth, because they understood we are but mere visitors on this planet.

I emigrated from Pakistan to the United States at the age of six. My extended family, and the two hundred million residents of

Pakistan, are at risk of becoming refugees by 2034. I live in Coney Island, an area of New York that faced the catastrophic effects of Hurricane Sandy. People lost their homes, businesses, cars, and yet, eight years later, the city still has not taken proper measures to amend the infrastructure of the island. There are many reasons why water pipes are still being installed, and streets are undergoing construction, but it's mainly because my community is low-income, Black, brown and immigrant. That is why I want to make sure that when policy for a liveable planet is implemented, it also ensures ecological equity in all sectors of the environment, from food to infrastructure to education in historically disenfranchised communities. Climate change is primarily influenced by the total stock of greenhouse gases (GHGs) in the atmosphere, and not by annual GHG emissions. Historically, developed countries and economies in transition have been responsible for about 75 per cent of the total global stock of GHGs. Pakistan contributes less than 1 per cent of the world's GHGs, yet its two hundred million people are among the world's most vulnerable victims of the growing consequences of climate change. According to the 2018 Global Climate Risk Index released by the public policy group Germanwatch, Pakistan will be among the first ten countries uninhabitable by 2040.

I joined the climate movement around October 2018. I started striking alongside Alexandria Villaseñor (a fourteen-year-old student who cuts school every Friday in protest, like Greta Thunberg). But it was awkward, because my walking out of college didn't accomplish much, except waste my parents' money. I attended XR training sessions around December 2018, learned about non-violent direct action, etc., and even helped organize an action targeting a US Immigration and Customs Enforcement (ICE) facility in Chelsea.

But I really started doing work around March 2019. The civil disobedience movement was lacking a key component: motivated and frustrated college students. So I started developing the software to reach my peers, and at the same time I was attending weekly meetings with the students in the climate coalition, who were working with Greta Thunberg's people to plan a massive march the weekend before the UN climate summit in New York. In the months leading up to the summit, I began giving talks at various universities in NYC. I am a rather reserved person, so mobilizing people required me to call upon capacities unnatural to me, or at least unpractised.

Nonetheless, I felt compelled to act, because the crisis at hand was bigger than myself. I talked about the science of ecological destruction, and my first-hand experience surrounding it. Unlike traditional classroom teaching, I had to convince students to take physical action after they walked out of our lesson. However, that was only a fraction of the operation. With the help of fellow activists, I developed a 'toolkit' that covered all procedural guidelines, from social media messaging, to recommended readings, to printable flyers and PowerPoint presentations on climate science and the history of activism, that students could use to build climate action chapters at their schools. And all of this work paid off, because on 20 September, thousands of college students walked out of their classrooms and marched from Foley Square to Bryant Park, and I had the privilege of leading them.

When we finally arrived at Bryant Park, my voice was at the brink of its limits, my abdomen had never felt so tight, and I was gasping for breath. Someone handed me a water bottle and I quenched my thirst while the rest of the team dispersed and went to prepare for the speeches they would give to the crowd of 315,000 people behind us. The sheer power of the echoes of the chants coming from the thousands of students behind me brought me to my knees, and I started tearing up. I would later find pictures of myself sitting cross-legged in front of a sign that read 'strike' trending all over social media. Never before in the history of mankind had millions of schoolkids, from all over the world, walked out of their schools demanding world leaders save life on Earth. In New York, we had walked down streets with nothing but megaphones in our hands, chanting 'Hey, hey, you, you, we deserve a future too!'

To give you an understanding of what 'organizing' this looks like, I am in a meeting almost once a day, every day of the week. A lot of the work is done via conference calls, emails and social media.

Each university is responsible for submitting a monthly progress report. Determine how much their school is taking in fossil fuel endowments. Where that money is going to. What net zero looks like for their school's business. Can it be done practically? If so, what infrastructural technologies need to be put in place, how much time would they need, and what is the price tag on those technologies? This gives the students an inventory of knowledge before they 'protest' so strikes cannot be broken by school officials.

That said, the climate movement is still brand new, and needs a strategic direction if it seeks to accomplish any major change. We got millions to show up on the streets and thousands agreed to get arrested, but we lacked a measurable demand. And the 'how'.

That's why I am now working with Indigenous activists Helena Gualinga and Isabella Fallahi, and already established movements from Madrid to Uganda, to come up with a plan to kick polluters out of conferences like the UN climate summit, COP, and, one by one, each land. It's an understatement to say this is a huge task, because the fossil fuel industry is the most corrupt and powerful industry on the face of the Earth. Which means we have to be as tactical, prepared and organized as they are.

DENIZ ÇEVIKUS, 12

Turkey

Deniz (born 2008) is a climate activist from the European side of Istanbul, Turkey. She works with Fridays For Future Turkey, and strikes each week in her city.

She has always loved sci-fi, horror and disaster movies, which inspired her first placard, 'EVERY DISASTER MOVIE STARTS WITH A SCIENTIST BEING IGNORED'. *She made that sign because she realized the link between global disaster scenarios in the movies and what could happen in real life. She still loves watching disaster movies, but now they also worry and scare her.*

Deniz loves animals, which is another reason she became a climate activist. She says, 'Animals contribute nothing to the climate crisis. They are totally innocent, but they are doomed to terrible sufferings because of it, because of us, actually. This really hurts me.' Injustice to animals is one strong motive for her amazing activism.

Inspired by Greta Thunberg, the founder of the FFF movement, Deniz started striking on the day of the first global climate strike in 2019. A day that was the starting point for many young people's activism, including my own. At first, Deniz went on strike in school. Then, she started taking strikes to the streets of her city. She's sometimes alone, but not always, often striking with other activists, such as her friend, Yağmur Ocak, who is also in this book.

Deniz is one of the younger activists in this anthology, but her age doesn't mean that she doesn't have something vital to say. Society needs to stop ignoring children like Deniz and listen to them instead. So read her words, and then act for the climate, because the time is now.

FACEBOOK: @deniz4future / INSTAGRAM: @deniz.cevikus / TWITTER: @CevikusHB

It all started for me when I heard about Greta Thunberg and her school strikes for climate. Under her influence, I delved into climate issues. By 2019, I had already developed a strong interest in the climate crisis. As I was observing that new strikers were showing up, from all over the world, each week, the idea of school-striking myself began to get into my mind. My decision to take actual action, however, came right after I saw a picture of a seal frozen to death. It was dead because of an extreme cold wave affecting its habitat, because of climate change, as we called it back then. I've been always very fond of animals. Injustice to animals moved me to deep sadness. When I learned that many species were in danger of extinction because of human activities on Earth, I decided that I could not take it any more.

Upon that decision, I did my first school strike for climate on 15 March 2019. It was the day of the very first global climate strike. At that time, we were only between two and three hundred people in Istanbul. I prepared my first two signs a few days before that strike: 'EVERY DISASTER MOVIE STARTS WITH A SCIENTIST BEING IGNORED' and 'WE ARE RUNNING OUT OF TIME'.

I joined the strike with my signs and marched with the other strikers, around the park. We were mostly children below thirteen years of age. Even some babies and three-year-old kids were among us that day!

Anyway, after the first climate strike, I was much too exhausted. But at the same time, I was happy as can be. It was the first time I 'did something' to save my own future. To save millions of animals' futures. To save billions of people's futures. To save the Earth…

I was so excited and hopeful back then. Later on, I would realize how hard it was to make people understand the awful situation we're all heading to, and also how hard it was to persist in striking for the climate, and being an activist. So, did I give up my hope?

No. I still have hope, but the more we lose time, the more my hope weakens.

After 15 March, I went on strike for climate every Friday, except for two weeks when I was too sick to get out of bed. Yet I wasn't striking on the street in the beginning. Until summer, I striked at my school. Sometimes a few of my classmates joined me. Till the end of the school year, I kept striking in our school garden.

Meanwhile, we did our second global climate strike, on 24 May. We were much more crowded than we were at the first one.

When the school year was over, Greta said school strikes would continue. The climate crisis wasn't going to wait for us! Even though it was time for summer vacation, strikers kept on striking, in groups or solo, raising awareness, demonstrating their determination, drawing attention to the persistence of the movement and making sure that it never falls off the agenda.

The summer of 2019 became the time when our voices began to get louder. Like many climate strikers around the world, I went on with my weekly strikes. I was striking at a different location every Friday. I also started to use a different sign. It was in my native language – Turkish – and it read 'ARE YOU AWARE OF [THE] CLIMATE CRISIS? I CAN TELL YOU IF YOU LIKE.' It was literally inviting people to come and ask me questions. My new sign attracted so much more attention than the previous ones. People were approaching me now, willing to hear what I had to say. As I read somewhere and noted down some time ago, 'Effective activism requires effective communication; in public and online.' That proved exactly true in my experience, too.

Then, the school year started again. I didn't go back to striking at school. I kept striking in public. The third global climate strike day was on 20 September 2019, and we did our biggest strike in Istanbul. There was something special about that strike: adults joined us. It was the most crowded climate strike in Turkey. We were 5,000 people. We met at a location on the Asian side of Istanbul, and marched down the streets to a park, under heavy rain, shouting slogans. At the park we made speeches, as climate activists of the city. The event proceeded with different activities till late in the evening, in spite of the weather.

And then again, I kept school-striking on Fridays. Since I believed in the influence of our weekly strikes, I never skipped a week and striked every Friday, even if I had to do it solo (but not alone). According to me, collective strikes and solo strikes complete each other. They are like two wings of a bird. Without one, the other does not work. A bird cannot fly without both wings.

My favourite strike location has always been the Eagle Statue, in the Beşiktaş district. I often go there to strike. It's a very busy location, where many people approach me to talk about climate. I like interacting with people. When I interact with people, I can feel that I'm being useful. Otherwise I could lose my courage one day. Doing something with no results would be disappointing and discouraging for anybody. I have to protect myself from that feeling. That feeling can make me stop. And I don't want to stop. I cannot stop.

To keep interacting, with young people especially, I'm visiting schools to make speeches or presentations on the climate crisis and FFF movement. I love meeting students and telling them about our burning future, encouraging them to be activists as well. They sometimes get afraid of the facts I'm introducing, but that fear will hopefully urge them to start acting for climate. At least I hope so. The world needs more climate activists and more school strikers.

In my opinion, the most important thing we can do at the moment is to talk to more and more people about the climate emergency, awaken them, point at the consequences, show the solutions, and urge them to get into immediate action. Because once they are aware of the danger we're facing, they can all be expected to act. From the outset, this has been my guiding principle, and also my motto in Turkish. That has been what I believed and aimed for all this time.

DEVANSH DESAI, 18

India

Devansh Desai (born 2001), nicknamed Dex Supertramp, is a climate activist from Gujarat, India. He has worked with the #SaveCongoRainforest campaign, striking daily for a while. He isn't part of Fridays For Future India, but he sometimes writes speeches for their strikes. He also picks up plastic and raises awareness online through making posters and graphics and tweeting about the extreme pollution in his country, as well as other issues such as the killing of tigers and other endangered animals. He has sometimes even made sure to record the levels of pollution each day, noting when it becomes worse.

Dex says he is 'attached to this [movement] because I love nature and I feel very bad... [when I] see that people... kill nature for their own benefit'. Another reason for his activism is that he sees how politicians are 'ruining the future for... [their] benefit'. And he also says that the Indian media do not talk enough about the climate crisis, instead preferring to focus on things like politics and religion.

As well as raising awareness through posting about issues like pollution, Devansh supports his fellow climate activists through cheering them on, and also makes computer graphics about the climate crisis. For instance, he has made artworks of several people in the #SaveCongoRainforest team, including Vanessa Nakate (whose essay appears later in this book) and me.

Knowing that climate education is important, Devansh chose to write about the science of the climate crisis – the reasons for it, the effects of it, and what the future could be like. This is a very important aspect of the fight against this crisis. Knowledge is totally necessary if we are going to change things.

In India, activism isn't very easy. Dex says that 'those who raise [their] voice[s] about climate change are not given much attention and are silenced'. But this doesn't stop him standing up for the climate.

TWITTER: @thewarrlock

CLIMATE CHANGE IS REAL.

'Climate change' refers to the change in the environmental conditions of the Earth. This happens due to many internal and external factors. Climatic change has become a global concern over the last few decades. Besides, these climatic changes affect life on the Earth in various ways. These climatic changes are having various impacts on the ecosystem and ecology. Due to these changes, a number of species of plants and animals have gone extinct.

WHEN DID IT START?

The climate started changing a long time ago due to human activities, but we came to know about it in the last century, when we started noticing climatic change and its effect on human life. We discovered that the Earth's temperature is rising due to a phenomenon called the greenhouse effect. The warming up of Earth's surface causes much ozone depletion, and affects our agriculture, water supply and transportation, and causes several other problems.

REASONS FOR CLIMATE CHANGE

Although there are hundreds of reasons for this climatic change, we are only going to discuss the natural and man-made (human) reasons.

1) NATURAL REASONS

These include volcanic eruption, solar radiation, tectonic plate movement and orbital variations. Due to these activities, the geographical condition of an area can become quite harmful for life to survive. Also, these activities raise the temperature of the Earth to a great extent, causing an imbalance in nature.

2) HUMAN REASONS

Humans, due to their need and greed, have done many activities that harm not only the environment but themselves too. Many plant and animal species go extinct due to human activity. Human activities that harm the climate include deforestation, using fossil fuels, industrial waste, different types of pollution, and many more. All these things damage the climate and ecosystem very badly. And many species of animals and birds became extinct, or are on the verge of extinction, due to hunting.

EFFECTS OF CLIMATIC CHANGE

These climatic changes have a negative impact on the environment. The ocean level is rising, glaciers are melting, CO_2 in the air is increasing, forests and wildlife are declining, and water life is also getting disturbed due to climatic changes. Apart from that, it is calculated that if this change keeps on going, then many species of plants and animals will go extinct. And there will be a heavy loss to the environment.

WHAT WILL THE FUTURE BE?

If we do not do anything and things continue to go on like right now, then a day in the future will come when humans will become extinct from the surface of the Earth. But if, instead of neglecting these problems, we start acting on them, we can save the Earth and our future.

Humans' mistakes have caused great damage to the climate and ecosystem. But it is not too late to start again and try to undo what we have done until now to damage the environment. And if every human starts contributing to the protection of the environment, then we can be sure of our existence in the future.

We have got a lot from this single planet, in this solar system, that is full of life. And we have lost our way and have forgotten that everything matters to Mother Earth. And it is our duty to protect Mother Earth from people who refuse to see the beauty.

Thank you.

Dex Supertramp, climate activist.

ELIZABETH WANJIRU WATHUTI, 24

Kenya

Elizabeth Wathuti (born 1995) is an environmentalist and climate activist from Nyeri. She is Head of Campaigns and the coordinator of the Daima Coalition – a coalition of civic actors who are advocating for the protection of urban green spaces – at the Wangari Maathai Foundation. In 2016, Elizabeth founded Green Generation Initiative, which aims to nurture young people to love nature and be environmentally conscious at a young age through tree growing, environmental education and greening schools. Through the Green Generation Initiative, Elizabeth also educates young people about the climate crisis.

Elizabeth has been passionate about the environment since the age of seven, when she planted her first tree. Having grown up in a place where there were beautiful forests and clean streams, Elizabeth had always loved nature, and so she decided to start reforesting her country.

Her work with the Green Generation Initiative earned her a prestigious scholarship and funding, with which she was able to further her studies and pursue her ecological aims. Elizabeth has also been named a Green Climate Fund youth champion and is a youth council member of International Reserva, a youth land trust. The Africa Youth Awards also named her one of the top 100 young influential Africans in 2019 due to her continued impact and influence in the environment and climate movement. As well as these things, Elizabeth attended the UN climate summit, COP25, and was a UN Young Champion of the Earth Regional Finalist for Africa in 2019.

Trees are vital to our environment. They absorb carbon and release oxygen, making them essential to our survival. Forests are also home to massive amounts of biodiversity, and to Indigenous communities. Therefore, reforestation is a hugely important part of climate action.

When floods take lives across her country, when the forests she loves are threatened or cut down, Elizabeth just fights harder. Because she knows that change has to happen, and it has to happen now. Africa as a whole is already suffering massively from the climate crisis, and Elizabeth is determined to make sure that no one forgets this.

If we stand together and act, things will change. When Elizabeth planted her first tree, she was just seven. Now, at twenty-four, she has helped thousands of children and young people to follow in her footsteps. And she did, does, and will continue to inspire people across Kenya, Africa and the world to stand up for our forests, and our planet, and to create a brighter, greener world.

INSTAGRAM: @lizwathuti / TWITTER: @lizwathuti

We all had our childhood moments. Whether bad or good, there are things that we loved, and held dearly to our hearts. Maybe some of these things shaped us into who we are today.

My name is Elizabeth Wathuti, I am an environment and climate activist from Kenya, and I am twenty-four years old. Having grown up in Nyeri, a region in Kenya known for its beautiful forests, and in a village where planting trees and drinking from clean streams was the norm, I got to love nature and connect to nature at a young age. My first act as a climate activist was planting my first tree, when I was seven years old, inspired by the late Nobel Laureate Professor Wangari Maathai, who at that time was our Member of Parliament in my home region. I became part of nature, and nature became part of me.

I believe that humanity's interaction with nature and our environment today has everything to do with how we have connected, or disconnected, with the natural environment. Nature has been my greatest teacher and my love for nature made me conscious of the environment at a young age. As a teenager, it always broke my heart when I saw, or read about, forests being destroyed or burned down, rivers flowing with plastic waste, animals being hunted to extinction, children struggling to breathe in some parts of the world due to how badly the air was being polluted. Everything was happening so fast and I was greatly disturbed and worried about how the future of our planet would look.

That is why I decided to get into environmental activism. All I want is for everyone to have a liveable world and a safe future, including all generations yet to come, and I will keep fighting.

In the year 2016, I founded the Green Generation Initiative, so that I could nurture more young people to love nature and be conscious of the environment at a young age. Despite being the most vulnerable to climate change, I believe children and young people in general have a great role to play in driving global action to address this crisis. That is why we cannot afford to leave them behind or ignore them.

On founding the Green Generation Initiative, I started out making baby steps. My passion was my driving force and I started without funds. My first event was in a primary school, where I used my own money to buy tree seedlings that would facilitate the tree planting and wider environmental education that day. It was a great success, and later on, due to my outstanding passion and personal commitment to environmental conservation, I received the Wangari Maathai Scholarship award and fund from the Green Belt Movement, the Kenya Community Development Foundation and the Rockefeller Foundation. This award enabled me to complete my dream course at the university, which was a bachelor's degree in environmental studies and community development. It also enabled me to establish my own tree nursery at home, so that it would be easier for me to source future tree seedlings at minimal or no cost.

Green Generation Initiative focuses on nurturing young environmental enthusiasts through promoting a nature-first culture, love for nature, and environmental consciousness. Our main programmes include practical environmental education, greening schools, inculcating a tree-growing culture among people to increase forest cover through an 'adopt a tree' campaign, and establishing food forests, where we incorporate fruit tree-growing for food security.

So far we have trained and nurtured over 20,000 children in different schools across Kenya to love nature and to be conscious of the environment at a young age, and of their role in addressing the ongoing climate crisis. We have facilitated the planting of over 30,000 tree seedlings in schools, where we promote the 'adopt a tree campaign', to ensure that the planted trees get to grow to maturity.

I do believe that everybody has a key role to play when it comes to addressing the climate crisis. We must also act fast enough, because it is no longer a future concern, it is happening every day. Climate change has always been a reality for us in Africa, because we are greatly feeling the impacts and they are affecting our day-to-day lives.

Many people say that Africa will be the hardest hit by the impacts of the climate crisis, but the reality is that Africa has already had to go through a lot of challenges, as the reality of the climate crisis has already hit home. It is not a future concern, it is about now, it is already happening and we are worried that this could only get worse for us, and yet we contribute the least to global emissions.

My advice to fellow climate activists is that we have to keep fighting, this is about our future and we have to secure it now, because we do not have time. As the impacts intensify over time, we, the young people and children of today, will face the worst effects and live longer with the consequences of the world's inaction, but we refuse to give up without a fight. We who see the urgency will continue to rise up, act, speak up and demand urgent climate action. We have to stand strong together and support each other from all corners of the world, both the Global South and the Global North.

We also need more climate activists. The more we are, the louder the voices and the more the impact.

ELLIOT CONNOR, 17

Australia

Elliot (born 2002) was born in Britain, but he lives in Sydney, Australia. This is yet another place where climate breakdown is evident, and where people are suffering, right now. Recently, Australia has seen terrifyingly high temperatures and catastrophic bush fires, which have affected thousands of people and driven koalas close to extinction.

Elliot participates in climate strikes in a low-key way, helping out with local organization and photography. Most of his work is focused on wildlife conservation. He contributes to twenty-four ENGOs (Environmental Non-Governmental Organizations) and directs his own, Human Nature Projects, which aims to reconnect people to the planet and remind us that we are in fact not separate from nature, but rather a part of it. This is now a major organization.

He was one of the youth ambassadors for CoalitionWILD in 2019, and this is just one of the many roles he plays in lots of different climate and conservation groups.

He has done all kinds of amazing things, including spending time in a castle in Southern France, caring for animals such as hedgehogs and an owl, and volunteering at all opportunities with organizations like the World Wildlife Fund (WWF) and the New South Wales National Parks and Wildlife Service.

He seeks to act as a 'voice for biodiversity, for the environment, [for] all those downtrodden, unrepresented lifeforms' which are suffering because of the current, privileged lifestyle of certain humans.

He is passionate about nature, and willingly takes on saving the world in every way he can, even though he is still at school. His story is a lesson to all those who say that they support us, but they simply don't have the time. Because if Elliot can find the time, you can too.

In his essay, Elliot discusses hope and solutions, and why Human Nature Projects began. The world may feel bleak right now, and yes, many things are *bleak. Yes, the future looks dark. But we can turn this around, with the help of people like Elliot.*

INSTAGRAM: @elliotconnor.eco / TWITTER: @eco_elliot

Counting Down to Crisis: Chasing Hope amidst the Climate Emergency

'Earth provides enough to satisfy every man's needs,
But not every man's greed.'

—MAHATMA GANDHI

I live in Sydney, Australia, where the temperatures in November 2019 were topping 35°C, the bush fire danger rating was at 'catastrophic' and hundreds of separate occurrences broke out – affecting thousands. Meanwhile I was travelling to Seoul for a conference on renewable energy and the need for better urban design to facilitate drawdown of carbon from our air. Coincidence? I think not. Because across the globe, millions are waking up to the climate catastrophe which now is unfolding. And our voices, so amplified, are being heard.

There's a whole rhetoric surrounding the activism–apathy divide, with political leaders sitting on one end and the likes of XR, FFF and Greenpeace on the other. I propose a simplification to the equation, whereby activism simply becomes a mindset, a description for all those who seek to redefine the Edifice of Human Nature in the present day. As with all things worth fighting for, it's not an easy change to make. We need to shift public dialogues away from egocentrism towards ecocentrism, and turn dim awareness into appreciation and respect for natural systems.

In January 2019 I was lodging in a castle in Southern France, freezing my fingers off caring for hedgehogs, bats and raptors, as well as teaching an injured owl how to fly. In the long, cold, dark winter nights I spent there, I did some thinking. Clearly what we were doing was insufficient, somehow lacking in the whole strategy of environmentalism. Two hundred species going extinct every day;

global temperatures smashing all-time highs – something had to give. And yet, given all these complications and competing issues, the solution which I envisioned was simple.

Under the status quo, humans present the greatest threat to humanity, and the greatest barrier to reform. That is to say, with the ecological IQ of society at an all-time low, there simply isn't a widespread understanding of the stakes we're playing with, less still of the solutions at hand. Hence by preaching beyond the choir, educating and engaging wider audiences in the wonder of nature, our current downward spiral might be turned around. Every one of us changes the world every day, but we don't stop to consider the change we're making. If people know and care about natural systems, that tremendous inertia can be used for good. And so, in June 2019, Human Nature Projects was formed.

With almost 2,000 volunteers across more than a hundred countries, our first steps have been a great success, providing a platform for everyone to enter into environmental pursuits. The secret to this success lies in the 4 C's of conservation: connection, curiosity, creativity and collaboration. Combined, these have the power to revolutionize people-powered preservation of natural spaces. Action brings about hope, and nature is helped to restore itself. It's that easy.

So ending on a lighter note than that on which I began, I would say to any aspiring leader that having a big, complicated problem doesn't necessitate a complex solution. Youth are the true leaders of today, but tomorrow is where our dreams can be realized. By linking up, reaching out, taking that leap of faith to start something without knowing of an end, we can collectively raise our voices and create impact. Remember: saving the world is like a baboon. You risk life and limb getting tangled up with it, but boy does it look attractive from behind!!

Elliot Connor – Founder and CEO, Human Nature Projects

EMMANUEL NIYOYABIKOZE, 24

Burundi

Emmanuel (born 1994) is a youth climate activist from Burundi, Africa. He is the founder and director of a reforesting organization called Greening Burundi. He also teaches other youth about the climate crisis and its impacts.

In Burundi, the effects of the climate emergency are already being felt, as Emmanuel explains. Warming causes the soil to become infertile, the agricultural sector is damaged, people starve. But we in Europe aren't hearing about what is happening there. So we need the voices of eco-warriors like Emmanuel to tell us about it. And we need to listen. I had no idea what was going on in Burundi. I can say truthfully that I think a lot of people in the UK haven't even heard *of Burundi. And I hope that Emmanuel's powerful essay can help to turn that around. To make sure that not only do we know that his country exists, but that it is suffering, and that we must change our ways to help it.*

Reforestation is a very important part of climate action. Trees absorb carbon emissions, and release oxygen. They are also havens for creatures such as birds, insects and apes. That's why Emmanuel's work is so important.

Poorer parts of the world, such as Burundi, are the places least responsible for global emissions, and yet they are the countries being hit the hardest by the consequences of climate breakdown. This is why we need to stand up for equity. This means that people in rich countries need to get to zero emissions much faster, to allow poorer countries to build some of the infrastructure we have already built. It means that the rich few who are mainly responsible for this crisis need to shoulder that responsibility. And it means that we fight the underlying toxic culture of valuing some lives above others.

Emmanuel has called me 'little sister'. But the people where I live aren't thinking about their sisters, their brothers, their siblings in countries like Burundi. This saddens me deeply, and I hope that this essay, and others like it, will turn that culture around, and turn our selfishness around.

INSTAGRAM: @niyoyabikozeemmanuel / TWITTER: @EmmanuelNiyoya

I'm a youth climate activist from the country of Burundi, a native of Bubanza Province.

I'm the founder and director of the Greening Burundi Project, which aims to plant millions of trees and reforest Burundi. My goal is to fight against climate change and its impacts that are happening in Burundi and the whole world. I consider myself an eco-warrior and dedicate my work and energy to making our world a better place to live. I feel concerned and see myself as a productive youth, embarked on reforestation in order to preserve the environment.

Following impacts such as deforestation, desertification and global warming in Burundi, the soil gradually becomes infertile. As a result, the agricultural sector, which provides a living for about 90 per cent of Burundians, is seriously damaged and production is deeply unsatisfactory. Poverty is lodged in households, diseases directly related to malnutrition increase, and the country becomes the first loser, faced with these plagues. As a result, I launched a project called Greening Burundi.

As one of the younger generation, I realized that we had to participate in empowering others to make Burundi better than it is today, when it is among the poorest countries in the world. I did not want to be a person who focused solely on my academic field, but also on how I had to spend my life as a productive member of the future generation. I am a kind of person who follows current affairs. I'm interested in what's going on in the world right now. What are the objectives of the United Nations? What do they want to achieve in 2030? What kind of problems does our world need to solve right now? Where is my position as an agent of change? Where is my position as part of the young generation that will continue to develop my country and our world? Am I doing something about it? These questions haunt me.

Our planet needs our attention, we must take care of it.

Our leaders must show a suitable example to everyone for the preservation of the environment by implementing actions, and avoiding the inaction which is among the causes aggravating the destruction of our planet.

Destroyers are invited to preserve our planet, change their behaviour and take care of the planet that is the lungs of everyone. Without Planet Earth preserved, life will be non-existent.

I invite them to not let their activities destroy our environment.

We young climate activists must continue our fight to make our world better and encourage people to take action for the climate.

FERNANDA BARROS, 15

Brazil

Fernanda (born 2004) is a teenage activist from Belém, Pará, Brazil. She joined Fridays For Future in January 2020, and started to strike, with her friend, outside the palace of her state's governor, with a sign reading 'Sim! Nós somos o futuro. Se houver futuro' – '*Yes! We are the future. If we have a future.*'

Fernanda is part of Aliança Pela Amazônia, the Amazonian branch of Fridays For Future. As she lives near the Amazon rainforest, a big part of Fernanda's activism revolves around protecting it. Despite it being the world's largest rainforest, and one of the lungs of our planet, not to mention the fact that it harbours huge amounts of biodiversity, the Amazon is being destroyed at an appalling rate, with disastrous consequences for Indigenous communities, animals and the planet.

Fernanda's activism was inspired by Vanessa Nakate, a Ugandan climate striker whose essay appears later in this book, and Greta Thunberg, as well as her brother, who was the person who first encouraged her to take action for the climate.

Although often striking with just one or two friends, Fernanda isn't discouraged. She knows that globally, millions stand with her. And she knows that most things start small, so she has confidence that more people will join her soon.

Another big motive for Fernanda's striking is the fact that Indigenous peoples are being so badly affected by the climate crisis, and have been for years and years. She knows that Indigenous rights cannot wait any longer, and that we MUST listen to Indigenous communities, because they have so much wisdom about the climate, about ecosystems, and about nature. Wisdom we have lost.

Yes, things start small. Yes, Fernanda is one person. But, as history has shown, one person can make a difference. And as we only need a certain percentage of the population to achieve system change, each person is a step towards the world we need.

INSTAGRAM: @nands_br / TWITTER: @Nands_Barros

Hello, my name is Fernanda Barros. I live in Pará's capital, in the north of Brazil. I am fifteen years old, and I am worried about the environment in which I live.

I entered Fridays For Future in January 2020, after some calls from my brother trying to convince me to start the movement here in Belém. I knew this was the right thing to do and about the importance of the cause. I was anxious, so I took some time to start doing something.

In the beginning, everything was okay, just me and some friends going to protest about climate issues. However, people around me started to ask me 'but is it only you?' It isn't a problem to me, but most people just want to join when more people are already doing it. But it is important to remember: you always start small, and the climate cannot wait, neither can Indigenous rights.

I went to the front of the palace of the governor of Pará, with my friend. We brought some signs and we stayed there showing them to the people around. I felt useful, doing something I trusted was right, doing what I could about climate change.

The current federal government incentivizes the climate emergency, and the destruction of the Amazon rainforest. Even with the wood market almost illegal here, the government still supports the exportation of natura wood (native wood from the Amazon), which is illegal, actually. The same issue happens with mining exploration on Indigenous land. The government support projects that increase the deforestation and cause even more Indigenous murders in the Amazon. It is absurd!

Everybody would win in a more sustainable world. We would have a beautiful Amazon, complete, rather than the equivalent of two football fields being cut down every minute. We would have Indigenous communities happy, the Amazon rivers without pollution, turtles without plastic straws, Venice's water transparent and clean.

The current situation in which the world finds itself is so alarming and depressing, and everyone is watching this, everyone is watching what only thinking about profit does. The world is hotter, people are already feeling prolonged droughts. So we need to change now, because climate change doesn't wait.

I'm very thankful to Vanessa Nakate and Greta Thunberg, for telling the truth about what is really happening in the world and opening the eyes of many young people, I'm thankful to my brother Abel Rodrigues, who called me and informed me about the group which, in Europe, is speaking about the Amazon and its peoples. Thanks for your daily struggle.

So if you want to, young activist, do something! Don't be shy, give your best, because we are needed, it doesn't matter if you are a beginner and people don't listen to you, and think that you are crazy, if you are doing this with only one or two friends, because you have international support and the world thanks you.

We know that the measures that have been taken are not enough, we see this on TV. We know what's causing everything, we have eyes and voices, and we can act.

FIONNUALA BRAUN, 19

Canada

A student of medical history, Fionnuala (born 2000) is the founder of the weekly climate strikes in Saskatoon, Saskatchewan, Canada, and the coordinator of Fridays For Future strikes in the prairie region. She also supports lots of other different climate campaigns.

Originally from Edmonton, Alberta, Fionnuala grew up spending a lot of her time with her dad, who is an entomologist and worked for the Canadian forestry services. Because of this, she learned about nature and insects, and how they play a crucial role in our ecosystems. Both her parents helped inspire her to fight for climate justice, and taught her to love nature. She says that some of her happiest childhood memories are of backpacking and hiking.

But there is a flip side to what sounds like a paradise. In Canada, the climate crisis is a definite reality, and she has grown up seeing its effects. She talks about how many of the places she used to visit aren't the same any more. But unlike many, Fionnuala didn't treat this as something inevitable. In fact, she found her way out of depression with action.

Alberta, where Fionnuala grew up, is home to oil sands, and therefore many of its people see climate activists as threatening to this source of income. But, conversely, there is also a community of those very activists who fight endlessly for a better world, and Fionnuala's mother is one of them.

Surrounded by two very different realities, where a huge source of the fossil fuels that created this crisis is not far away but right there in her state, and yet also where activists are determined to change things for the better, Fionnuala has insight into the climate crisis in ways that many of us do not.

Striking every week, no matter what the weather, continuing bravely onwards with determination... Fionnuala is proof that there are tons of amazing activists who don't get the recognition they deserve.

INSTAGRAM: @fionnuala.sb / TWITTER: @fionnualaAnjoo

What is the climate crisis? Is it something we hear about on the news, consider briefly and feel some mild concern about, and then turn our backs on and continue our daily lives? Is it something we turn our backs on because we are afraid of the implications of it, and hope that by ignoring it, the crisis will disappear? As a young Canadian activist, neither of these takes has been my way of thinking. However, I can understand why people, especially those who come from specific demographics or locations, would feel this way. The climate crisis is an enormous issue, an all-encompassing topic that requires a massive overhaul to the very foundations of our society should we hope to reverse its effects. It's enough to make any seasoned politician cringe, let alone an ordinary citizen. However, in Canada, we are in many ways on the front line of both the environmental and economic fallout of the climate crisis. Choosing to ignore the ecological breakdown of this planet is no longer an option for us. But if there is one thing I have learned from being an activist, it is that you are not as alone as you may think.

Canada is considered one of the most naturally precious and breathtaking places on the planet. I grew up in Edmonton, Alberta, a four-hour drive from both the sweeping slopes and awe-inspiring glaciers of the Rocky Mountains, and the scarred and pitted earth that marks Alberta's leading source of income: the oil sands. In many ways, this dichotomy is representative of Alberta's mindset, a strange melee of hardened oil sands workers, made wealthy and stable because of their jobs in the North.

However, there is also a deeply rooted community of climate justice activists. They work day and night to meet the Paris Agreement targets, working with the provincial government to encourage green energy in homes, and working with oil sands workers to reassure them that there are career options available for them in a green economy. My mum is one of these activists, and in many ways, she was the one who taught me about climate breakdown and showed me how to take action and make a difference. I can remember travelling with my parents to Jasper when I was young, visiting the Edith Cavell glacier, and hearing my mum explain to me that in twenty years, it could all be gone. I can remember the trees that filled the valley surrounding the Old Fort Point and Valley of the Five Lakes hikes, in which we would walk together as a family every fall. Now, the Edith Cavell glacier is nowhere close to the size it was when I first visited, and the trees that used to provide shade on the many fall hikes we took are gone, cut down due to the increasing migration of the mountain pine beetle.

And yet none of this devastation inflicted on the world-famous Rocky Mountains is even close to the other effects of climate change that Canadians, as a whole, have begun to experience as the climate crisis continues to progress. More extreme weather (in Saskatoon, where I live now, the temperature reached -50°C in winter 2015 – it was nothing short of insanity), as well as more frequent and intense forest fires and flooding, causes hundreds of thousands of dollars in damages to middle-class families every year. Every year, more western Canadians lose their homes to forest fires than ever before.

Indigenous populations are also disproportionately affected; as more species go extinct, not only are these communities placed in peril in terms of being able to hunt and provide for themselves, but hunting and gathering are essential to the Indigenous identity and way of life. However, Indigenous communities, as well as many other groups demanding climate action, feel unheard by the Canadian government, which continues to invest in dangerous pipelines, encourage oil extraction and dance around the real question of a plan to address climate change. As one of the primary regions affected both socially and economically by the climate crisis, many activists feel as though they are fighting a losing battle against governments and corporations.

For myself, the climate crisis and the activism it has inspired in me does not just have to do with the fear I feel about the ecological breakdown of our planet, although that is undoubtedly one aspect of my experience; climate activism has also inspired positive change in my life. I have always felt a deep need to effect change. When I was small, I can remember wanting to get involved in everything, every after-school club, student body government and student panel. However, as I got older, I struggled with being too shy, finding myself unwilling or unable to speak my mind. My self-esteem plummeted, I began to have difficulty eating, and I struggled to function within social settings.

This all changed for me, however, when I saw another young woman, one who experienced similar struggles to me, taking a stand. It reminded me of how I was always raised to fight for my planet. It reminded me of the fear and hopelessness, which had morphed into apathy, that I felt about the climate crisis. And so I began to act. Simple acts like spreading the message that we should be listening to the science, striking every week and becoming involved in local organizations have instilled me with new energy. This is why I encourage every young person to act, to become mobilized, and to fight for their future.

Of course, there is also the less empowering side of the fight against climate change. Every week, when I go out to strike, I can't help but feel a deep, visceral fear for my future. It makes me sick to my stomach to know that children effectively have to beg politicians for their lives, and yet still very little has changed. When I sit down to write a speech or draft a letter to the government, I often find myself frustrated because I feel like everything I have to say has already been said. Over and over again, the youth take a stand for their future, and it feels like over and over again, we are given empty promises and empty words. Politicians smile to our faces and sing our praises, tell us we are 'inspirational' and 'the leaders of the future'. But when I hear them say this, all I can think is that if they do not act now, there will be no future for us to lead. This brings me to my message to all leaders of today.

The youth should not have to be your inspiration. We should be in school, dreaming of a future in which we can live full lives, have children and bring innovations and ideas to society. The young people are not your leaders. We should not have to take the hand of the adults and drag them towards the solutions that will save our planet. We are still students, still learning how we fit in the world. We should not have to teach you how to lead. Therefore, our message is simple. Listen to the science. If you feel uncomfortable listening to the youth, at least listen to the people who studied for years who are telling you we are ten years away from the point of no return. Read the Intergovernmental Panel on Climate Change (IPCC) reports, and listen to scientists begging for change instead of slashing their funding and effectively silencing them.

And then, after you have heard the science, we ask that you act. Do not act to win more votes or more donations. Do not spread empty promises and lies. Inform yourself about the changes that need to be made, and then make them – no more holding back, no more pandering to the climate change deniers.

As I posited at the beginning of this piece, we no longer have the luxury of ignoring the crisis, of pretending it doesn't exist and hoping that someone will come along and fix everything for us. The climate crisis is terrifying. No one has ever denied it. But the planet is running out of time, and as long as you continue to betray us, the young people will continue to fight.

FLYNN, 16

England

Having an environmentally conscious vet for a mother was a good starting point for an activist in the making. Growing up wearing second-hand clothes and living in a house full of animals, including hedgehogs, deer and pigeons, Flynn (born 2004) knew something about the state of the climate. But it wasn't until the emergence of the youth strike movement, Extinction Rebellion Youth, and then the Stop HS2 campaign – where they, at the time of writing, live in the woods and are involved in direct action – that they truly threw themself into activism.

Flynn, and their friends, used to strike most weeks with Fridays For Future, and as well as spending most of their time resisting HS2 (the UK's largest infrastructure project) they also work with Extinction Rebellion Youth, UK Student Climate Network, Radical Restart and other environmental and social justice groups at a global, national and local level.

Despite being in the UK, where protesting laws are fairly relaxed at the moment, in the coming years the police may start to crack down, and after being arrested while doing a lock-on with their friend, causing some of HS2 to stop works for the day, Flynn is very interested in this and is starting to read in their spare time about policing alternatives and how the current system works (or rather, doesn't).

Living in the UK, Flynn knows they are privileged, but they are still seeing the effects of the climate and ecological emergency. Having asthma makes them increasingly terrified of the levels of air pollution in Oxford. Air pollution has direct links to respiratory diseases, and having asthma makes these a whole lot worse. Not to mention that pollution is directly linked to the worsening of asthma itself.

In October 2019, I met Flynn while doing actions in London. We both took part in marches and protests, and fought for our futures together. I was honoured to fight alongside such an amazing person, and am equally honoured to share their words now.

Although they label themself as 'one of many' climate and environmental activists, Flynn, like anyone, has a unique story to tell. A unique journey to activism. A unique perspective on the struggles facing our world.

INSTAGRAM: @flynn_.t / TWITTER: @anarchistcrocs / FACEBOOK: @Flynn T Upton

I'm sorry for starting off in a cringy way. But what even is an activist? I find that word so weird. You see, this whole situation is weird and scary. And sometimes it feels like us young people have to tell the rest of the world what is happening around them and that older people are either blatantly ignorant or constantly apologizing for not doing enough. Stop the platitudes and do something.

There are so many different things you can do, whether it is something big or something small. From taking part in direct action to emailing your MP, from living in a community to sharing information online about the climate and ecological crisis. Of course we need systemic change, but that begins in realizing our individual powers.

Another thing we must all do is recognize the privilege we hold and think about what that means for us on this Earth. I am privileged in lots of ways, especially as I have – so far – not been affected in such an intense way by the climate crisis, as many others have been. But I will be affected, just like everyone else. It is just a matter of time. The idea that rich people will be able to 'buy their way out of this' is another problem. You cannot buy your way out of the climate crisis, as it will affect everyone in one way or another. You can't run away from it either, so you may as well at least acknowledge it. Indigenous communities and people in the Global South are being affected right now, and have been for a long time. Yet it takes a white European girl in the media for people to open their eyes to this crisis. Think about that for a second.

Recently Oxford has had harsh weather conditions (hotter summers, etc.), which will just get worse, and we are becoming more prone to flooding, threatening our natural areas, which should be thriving. But that is not all. Very soon, air pollution within Oxford will worsen, and the NHS will become increasingly more overwhelmed. Are we prepared for this? I don't think we are. Considering we have seen the effects of immense pressure put on our healthcare system from the Covid-19 pandemic, it is completely understandable how I, an asthmatic sixteen-year-old, find this looming crisis terrifying and am scared for my future.

In the UK we are still able to explore and wander through many ancient woodlands. But more and more are being cut down every day to make way for a train line from London to Birmingham, which is only 'Phase 1'. We already have a train line running between the two cities that only takes one and a half hours. I have walked from Birmingham to London and in that time passed so many overgrown railway lines, which leaves me to wonder: are these fellings of trees and land grabs really for a train line, or is it for something else? A massive 92 per cent of the countryside and 97 per cent of rivers are not publicly owned, and the number of natural spaces grabbed by HS2 is only making precious spaces like ancient woodlands less available to the public.

The money being put towards this project is growing every day, and by the time you read this, it will have grown by at least three times (if not more). This vast amount of money should be put towards helping us fight the climate crisis: it should be towards health services, funding education, making sure people can eat and drink, and that everyone has somewhere to live. These are basic human needs. This insane project will only benefit the privileged and doesn't even benefit the local people whose back gardens, wild spaces and surrounding ancient woodlands they have spent all of their time in are now being destroyed, and what for? A journey that is only a few minutes shorter? This is not right – nature is a human right that the UK government continues to deprive us of.

A big part of what keeps me active in this is the fact that I am surrounded by such a lovely community. If I didn't have that, I would find it very hard to carry on. Once again, this shows the privilege which I am lucky to have, as many activists are on their own in this fight.

There is so much you can do to resist HS2. Wherever you are and whoever you are, if you are able to get down to the camps and stay for any amount of time, you would be most welcome. If you can bring food to the camps and take away rubbish, if you can share lifts or support arrested protectors outside police stations, it will be appreciated. When you have been arrested for slowing down HS2, it is so lovely to know that people will be outside waiting for you to come out. And if you are a local, please speak to us about how we can connect with the locals, and if you can offer up your washing machine for dirty clothes and tea towels that would be great! Learn to climb trees and cook and have interesting discussions with those resisting HS2. Unlearn the harmful things you have learned and teach yourself what you want to learn. There are many things you can do from elsewhere, such as emailing your MP and finding interesting pieces of information to share among people and social media. Mobilize people online, and do give whatever you can.

But why should we care so much? We could sit around and let them cut down trees. Trees which allow us to breathe. But why would we do that? Well, perhaps you are out there resisting it right now and maybe you can't and that is totally understandable. Some people are unable to for many reasons and like I said, I recognize the privilege I hold and use it wisely. Things get in the way. Life gets in the way. That is how capitalism works, and I hope you see the bigger picture. This is a representation of how the UK government is dealing with the climate crisis – chasing a dream of continuing economic growth at the expense of our natural wildlife. In other words, they are not going in the right direction.

I know it is hard, because the facts are scary, but they are real, and they are not going away. In fact, they are becoming worse. When you are in these communities and spaces just listen to the people around you. You can sense their emotions, and you will notice that we are all scared but have accepted the truth. Scared for ourselves, for you, for the people you know, for the people around them, for their kids. We are scared for our lives, and the lives of everyone on this Earth. It may seem like we have all the time in the world, but we do not. We are on a countdown, and our time is almost up. I am beginning to lose hope that something will happen, and that the people who can make changes will do so.

I hope that you will read my words, understand that I am writing with all my passion. Something will change, and at least I can say to the future, whatever form it may take, that I tried. I hope that you will be able to say the same.

Drastic system change is what we need to survive. And we need it now.

GRACE YANG, 15

USA

Grace Yang (born 2004) is an Asian American climate activist living in Massachusetts, USA. She works with the Sunrise Movement, a large youth movement in the USA, as a hub coordinator, and she is the co-leader of Fridays For Future Massachusetts. Grace takes a variety of other roles as well within FFF. She has also worked with Youth Climate Strike.

As well as helping with and taking part in big strikes, Grace silent strikes every week, sitting down and refusing to speak for a while. This can be a super effective form of protest, and carries its own message. As Grace says, 'Adults are not listening, so I am not speaking.'

After the founder of 350.org spoke at her school in 2019, Grace knew she had to do something. Afraid and angry about what was coming, Grace decided to organize a strike in her home town on 20 September that year, the day that, for many countries, was that of a huge global strike.

Grace is incredibly devoted, to the point that she knows from personal experience what burnout, stress and depression feel like. It can be so hard to step back, step away. Grace describes feeling like she had to do everything she possibly could, keep going whatever, as though everything would collapse if she didn't.

Letting go of this can be hard. Taking a break sometimes feels like a failure. But it isn't. Looking after yourself is super important. So this is what Grace Yang is reminding everyone. It's okay to take a break. More than okay, it's brave to recognize when you can't do something, and when you have to look after yourself. Brave to do so, and then, when you can, start fighting again.

Another thing Grace isn't afraid to do is speak up when there are things going wrong within the climate movement. Because we aren't perfect, and we do have to acknowledge that. Being an awesome activist isn't just about marches and protests. It's about making sure to speak out when things are going wrong within the movement. It's about making sure to stay true to your values. It's about taking breaks when you need them, because if you look after yourself, then you will be able to fight twice as hard.

INSTAGRAM: @withgraceyang / TWITTER: @withgraceyang

I wake up at 6.45 a.m. and, still bleary-eyed, I stomp downstairs to thirty Slack notifications that I respond to quickly, shoving away a little wave of stress. Throughout the school day, I obsessively check Slack under my desk during class, and check it right before I go to bed.

I'm Grace Yang, an Asian American climate activist and organizer. I have worked with Fridays For Future, the Sunrise Movement and Youth Climate Strike, as well as many other groups. You may have read articles about how climate activism improves mental health, and everything they say about finding community and purpose is true. But it can easily do the opposite.

I think that if I hadn't joined the climate movement, I'd have lost it. Conference calls keep me going, giving me a sense of purpose and forcing me to keep moving. There were times where I was crying, feeling like everything was falling apart, but having things in my schedule gave me a sense of structure and I would calm down, before getting on a call and feeling less alone. I feel less helpless about the world crumbling around me when I'm working towards a better future. I've learned so much from organizing, and being around amazing, smart, talented activists I'm proud to call my friends. The sense of community has gotten me through a lot.

Still, there's a different side to climate activism. I've felt like I needed to be online all the time, or else things would crumble and burn when I wasn't watching. I've struggled in school, because I was so busy worrying about what was happening in the movement, or checking my phone instead of paying attention in class. Drama plagues the movement, and we spend hours upon hours arguing online and on toxic conference calls instead of working to make real change.

The youth climate movement needs to remember its original purpose: to fight for climate justice. When we let ego get in the way, drama follows, leaving many drained and stressed, and taking energy away from organizing. We're here to stop the world from burning. We don't have time to waste.

I've simply taken on too much – I've been in countless organizations, and the workload on top of school can be brutal. I've avoided opening my inbox, or my to-do lists, to ignore the overwhelming amount of tasks.

My schedule on weekdays often looked like: get home, recover from an exhausting school day, respond to a heap of emails and messages, hop on a call or two, eat dinner, attend some more calls, then start my homework very late at night. I'm an overachiever, and I know this is not unique to the climate movement.

To all my lovely organizers out there: remember yourself. Don't sacrifice your sanity for the movement. I promise you, the movement will be better off if you get off Slack and Telegram and go to bed.

And that's my advice for other activists out there – don't let it take over your life. When I started giving up my other hobbies and passions to fill that space with activism, my mental and physical health dipped. It's okay to have fun. It's okay to have a life.

You'll still want to do a lot. And you should decide for yourself how much you want to take on. Relying heavily on Google Calendar has been a lifesaver for me, as well as making small to-do lists for myself and giving myself little rewards when I cross something off. Remember that many things will seem urgent, but you don't need to do everything right this minute. The world of activism can wait while you close your laptop for a bit.

HELEN, 21

Scotland / England

TRIGGER WARNING:
REFERENCE TO DISORDERED EATING

Helen (born 1998) is a British student studying history in Glasgow, Scotland, and an activist with Scottish Youth Climate Strike (SYCS) and the #SaveCongoRainforest campaign. She helps organize the strikes in Scotland, and, apart from that, strikes daily for the rainforest and weekly, often alone, for Fridays For Future. When she is in England, in the holidays, she does her activism there.

Throughout much of her life, Helen has experienced disordered eating patterns. In her essay, she reflects on the parallels between what she was doing to her body and what the leaders are doing to the planet. They have to make the choice Helen did. To put life first. To stop lying to themselves and to everyone else.

Helen has also struggled with anxiety, both GAD (Generalized Anxiety Disorder) and social anxiety, but, she says, 'The method I found most effective in combating these issues is activism.' Despite all her personal struggles, Helen has kept going with activism, but isn't afraid to take a break if she needs to. A vital, hard-working member of the #SaveCongoRainforest group, Helen is helping campaigns, but also her friends. Being there is such an important thing in this uncertain world, and Helen makes sure she is.

Helen says, 'If I ever go back to my "normal" life again, I would like to pursue my interest in history at a postgraduate level, but as it stands, fighting for a future is, right now, more important than studying the past.' So many young people are having to put their dreams on hold to fight for things that are basic human rights, such as clean air. And even though, living in Britain, Helen hasn't been hit too hard by the physical effects of the climate crisis, she knows that, if action isn't taken, she will be. And that so many people already are being.

Helen says, 'I never thought I'd be an activist, but I'm so glad that I am.' And given that we're in this situation, I am so glad, too, that we have amazing people like Helen onside in the fight.

INSTAGRAM: @carbontiptoes / TWITTER: @_carbontiptoes

This essay will not include numbers or weights, but if you are sensitive to discussions surrounding eating disorders, then please exercise caution before reading this contribution. – Helen

For a long time, my life has been discoloured by disordered eating. Many people involved with climate activism, especially young people, have reported an amelioration in their mental health struggles after joining the climate movement. I would like to be able to report that the simple act of becoming a climate activist made everything better. In reality, however, this was far from the case. Climate activism is not a cure for any climate-exacerbated mental health issues the only cure for climate-related mental health issues is for our governments to listen to the science, and act on the climate and ecological emergency. However, what began to happen as I became more involved in the fight for climate justice was that, everywhere I looked, I was being forced to acknowledge my own hypocrisy. How could I expend so much energy demanding that the government listen to the science while my own actions paralleled theirs, albeit on a much more personal scale?

Slowly, I started to see my own warped sense of reality mimicked in the way that our governments were ignoring the climate crisis. It was almost surreal; my own compensatory behaviours were being reflected back at me on a global scale.

Climate accounting was no different to the way that I would twist my calorie intake to pretend that I wasn't sick. I would joke about loving food in front of people to reassure them, in the same way that companies greenwash to shield themselves from their massive ecological impact. As I watched handfuls of my hair snake down the plughole in the shower, I would pretend that it was just the stress of university – while at the same time, governments everywhere refuse to acknowledge the environmental destruction that their policies condone.

When my lips became so dry that I couldn't smile without them cracking open and bleeding, when I'd shiver against my hot-water bottle in bed at night, when I dreaded standing up because I knew that my whole world would slide away into nauseating darkness, I would remind myself of the goal: thinness. I was taking everything from my body and giving it nothing in return, because my health did not matter an iota to me. Even when I was diagnosed in 2018, very little changed; becoming thinner and thinner remained more important to me than anything else. Our governments are treating our planet the same way. Their refusal to act on the science is a symptom of being blinded by an impossible goal; becoming richer and richer is more important to our governments than anything else.

Sometimes, it takes a sick mind to recognize a sick system.

I am under no illusion now that I was killing myself. This terrifies me, because I can see the impact that the same reasoning, from our governments, is having on the planet. However, I do have hope. I realized that I had to change, that my behaviour was unsustainable. I had to let go of the thing that was most important to me – losing weight. This was a wrench. It was difficult and scary and uncertain. I was losing my control, my way of life, throughout most of my youth, and I did not know what sort of world I would be facing when I recalibrated my priorities. However, because I finally accepted reality, I will, I hope, be alive and healthy in twenty years. I chose to live, rather than to simply exist.

We know that the endless economic growth of a fossil fuel economy is not compatible with survival of life as we understand it. We know that we must seriously lower our emissions in order to achieve the 1.5-degree limit of warming set out in the Paris Agreement, and we know that we need to act now in order to stand a chance. We have all the facts that we need. We now need the political will to change the system.

To our politicians, our leaders, I say this: change is scary. A total recalibration of our aims and motivations seems unnerving and impossible in equal measures. Regardless of how it may feel, you must realize that, sometimes, we have to be brave. We have to take that first step. Now, right now, we need you to take it for us. You have to listen to the science. You have to realize that no amount of money is worth the destruction of the planet. You have to realize that monetary wealth can no longer be the primary measure of worth.

You have to choose to let the planet live.

HOWEY OU, 17

China

Howey Ou (born 2002) is a vegan, zero-waster and climate activist from Guilin, China. She had already given up animal products and started educating people she knew about veganism and plastic, as well as giving speeches at her school, when she heard about Greta Thunberg and decided to start a strike, which, as you will read in her essay, didn't really go as planned.

Howey didn't know what to do. She grew depressed and her relationship with her parents deteriorated, until she left home and went on an environmental pilgrimage.

She became frustrated at the lack of climate awareness, even among NGOs and scientists. Then she received an acceptance to go to a UN Youth Climate Summit. When she returned home, she started to act once more, launching the initiative #PlantForSurvival, a movement centred around the importance of plants and trees, to offer a more widespread and accessible way for Chinese people to join the climate movement. Now the movement has gained traction, and more and more people are in contact with Howey and #PlantForSurvival. She also educates other schoolchildren about the climate crisis, raising awareness among her peers by telling them about Greta, her strikes, and the emergency we are facing.

Howey loves nature, and tries to reconnect with the natural world whenever she can, and encourages others to do it too. Her message to those who want to join climate activism is: 'Start with something small, like talking to your friends... It's [also]... important to really feel nature, you could do this by meditation... and walk in nature without anybody or anything other interfering [with] you. Last but not least, share love and support to everyone you meet, [in] every word you speak.'

Howey's story proves that if you have courage, determination and the knowledge that what you are doing is right, nothing and no one can knock you down.

INSTAGRAM: @howey_ou / TWITTER: @howey_ou

(EDITOR'S NOTE: Howey has told me that, for their safety, some proper names of people mentioned in this essay have been changed.)

Connect to Nature

I think my involvement in taking climate action was my destiny – it was something that began with an unexpected personal revelation.

During a winter holiday in February 2017, a brochure about veganism changed my future. It was the first time I knew about veganism, as in China, the rate of vegans is only less than 0.4 per cent. But this small book about global warming and animal rights really determined my mind to be a vegan. In the following year, I insisted countless times in front of different relatives, compromising, debating, fighting and crying with painful experiences, but I still ate fish from time to time. On 17 January 2018, the supernatural being just led me to the place I belonged. It's a dream which enlightened and saved me, leading me to the place I am now in understanding my relationship with the natural world.

In the dream, we were taken to a restaurant and instructed to kill one of the living fish. In the dream I vividly remember an image of myself holding a knife, being told I must kill the fish or I would not eat that day. I tried to grab the fish with my left hand, lifting up the knife with my right hand several times to slay it, with shaking hands and heart. The fish escaped from my evil hand with all its strength! One of its panic-stricken eyes made contact with my own. I felt the extreme eagerness of life, and extreme horror of death from a dying being. Its life depended on whether I killed it or not. The contrast even intensified the conflict with my own embarrassment. With the complexities at tipping point, all gathering to a horrible small black dot, stress at its maximum, suddenly it blasted.

Something gushed out of it like a flood. I woke up, with knitted brows and beating heart. The stressful atmosphere faded away slowly and hazily. I felt the hardness of the bed, and thought that what had happened just now was incredibly real. As I lay thinking about the dream, it was then I decided I did not want to consciously, ever again, be the cause of another creature's death, through the way I live my life.

However, I considered that I was only an observer instead of a real change-maker, until July 2018. I found a magazine called *National Geographic*, which had a special feature on plastic pollution. Soon I realized that it was the most important issue that I needed to concern myself about. So I shared the knowledge with my parents, like I was sharing new ideas, as usual. Then, I was inspired by my own behaviours. So I advocated reducing plastic to everyone I met, talking to each one for about half an hour. The more times I repeated it, the deeper I understood the issue. I became a real environmentalist and change-maker; I gave speeches in front of class and the school club. However, the gap between real life (plastics, plastics, plastics) and my dreams of blue oceans was so large that it seemed to be impossible to stop the disaster, so I fell into depression and anxiety. I could not even have a normal life and study like others, due to an involuntary anxiety every ten minutes, which lasted for half a year.

The experience I described above really gave me a unique life. I formed a much tighter connection with nature, which most people lack. And this made me have the bravery to do what others didn't do. I know, sometimes, if I don't do the right things, then nobody will do them. So I must keep standing up regardless.

In the first half of 2019, I gradually heard about a girl called Greta Thunberg, and read the news about her. Inspired by her, I started my first #FridaysForFuture protest in front of the Guilin People's Government in May 2019, as the first striker in China. I was arrested by the police after seven days protesting. I didn't know what I could do to continue non-violent direct action in China. Maybe because I was not so brave, I felt frustrated. I argued with my parents every day, once for eight hours.

Feeling frustrated and suffering, I was wavering, and left home with fourteen-year-old animal rights activist Nlocy. In two and a half months, I kept trying to find a way to develop non-violent direct climate action, and I travelled around China alone for much of this time. During this hard time, I lived in any place I could find, because my parents cut my living expenses. I got no money to eat and sleep, I slept in the hall of the hostel and on the bench of the university. I ate what others left and had to make friends and inspire them, so I didn't have to sleep on the streets. I went to all the activities that I could find related to the climate crisis and sustainability, to ask different kinds of people about their opinions and suggestions about movements in China, like B Corps, NGOs, climate scientists, permaculture labs, etc. Once I got a fever and headache at 11 p.m. when I didn't have a place to stay. I really knew what hunger and tiredness were. During this one person's pilgrimage, I felt really lost and frustrated, but also gained some experience about practising activism in China.

These following passages are about how I reconnect to nature now. On a day when I reconnected to nature after two months' quarantine, I sat down on the cobblestone road and wrote down all this.

'Stopping connecting to people and nature for two months especially made me more depressed and isolated than before, maybe even more egotistical. Now I am assured that connections in the real world are what could never be replaced by the internet. REALLY. I have been isolated from my other relatives, my familiar living conditions for more than one and a half years now. The density and noise from urban housing cut me off from nature. Now I can understand why Nanda (an eighteen-year-old vegan meditator and founder of a homeschooling platform in China) said that he always went to the nearby mountains, sometimes every day. It's the day-by-day call that helps one through the ordeal and cleanses one's soul. Surrounded by nature, gaining universal spirits and sharing love to every being from your heart are natural, and it's easier to go into higher dimensions and receive deeper thought. Maybe I can understand Zhang Qingyi's criticism about Hong Kong's high prosperity now. It's hard to say that why I took action is not because of my strong connection to nature now. WOW.

'I remembered when Hazel (the founder of a large veganism platform on WeChat, a Chinese social networking platform) gave speeches on VeggieWorld. My attention was drawn to Hazel's content about reconnecting to nature and old Chinese sayings. I was attracted to it and thought, "It's just describing me so accurately." It's just the reason why I care so much about the ocean's destiny, because the ocean is such a symbol of freedom and wilderness with millions of beautiful spirits and scenes in it. I now realized I loved those traditional cultures, they connected to nature so tightly and spread warmth to people, forming communities. I recall that in recent months, I didn't use electronics very much, even felt some difficulty when I picked them up. We NEED to recall, we need regenerative culture, really.

'Why am I so enraged about outrages to nature? Because I connect to nature. Nature relieves me, accepts me and inspires me, so generous, so living. Nature's pain is my pain, it is engraved in my heart so deeply that I sometimes don't even feel it. Nature and I have already mixed, become uniform. We are one already. I am lucky that I was born in Guilin, a beautiful city, with clear water and green mountains. Although it has grey sky and a plain life, I love the instinct that she gave me to love nature. Maybe she seems boring, maybe she seems plain, doesn't seem like Shanghai or Guangzhou, the metropolises that have creative atmospheres and international prosperity. But now, I think that it's just humans' ignorance and craziness that are cutting ourselves off from our Mother Nature. I even wasn't aware that Guilin had injected nature's spirits into my bones and blood from the time I was born. She pardoned my ignorance of her beauty in the past until just now. She gave me the chance to live in a university with luxuriant trees and grass, as well as those "socially distanced" undergraduates. It is a pure land that is including me gently, quietly, and has made my family flourish for decades. It is where I came from. Nature is my forever church.

'I feel the fire burning in my chest while I am writing all of this, full of ardour and love. It's a very valuable experience to write down my feelings so naturally and fluently. I am accepted by myself; I know myself; I know who I am, I know where I am, I know what I love. Thank you! I love you.

'I even don't wanna stop writing. I am unique. Nature is my church.'

KAOSSARA SANI, 25

Togo

Kaossara 'Kao' Sani (born 1994) is a climate activist from Togo, West Africa, with a Togolese American father and a Nigerian mother. She's a striker, educator and member of the Rise Up Movement Togo, the Togolese branch of an African climate action group which was founded by Vanessa Nakate from Uganda (whose essay appears later in this book). She also founded Climate Voices for Africa, a campaign that promotes African voices, because, as Kao is trying to get the world to see, AFRICAN VOICES MATTER!

The person who inspired her to act for the climate was Wangari Maathai, a Kenyan activist, tree planter and utter hero. Kao says, 'I remember the day I read her story... I was fourteen at the time... [and it]... fascinated and moved me. I said to myself, "yes, I would like to be like this woman. Strong, determined and fighting." [Others who have inspired me have been]... people I've met on Twitter. They are wonderful, united, and tireless.'

She is a sociologist, working with an NGO *that helps homeless children and gives them opportunities to go to school, as well as doing all sorts of things to help people in Togo help themselves, helping them both financially and by giving them her time and care. She also teaches many younger people about the climate crisis, going into classrooms to raise awareness of the impacts that it is having on our lives, telling them about such issues as deforestation, plastic and how we need global justice. She has raised awareness in at least six schools (and counting) that have since banned plastic. Plus, thousands of students will help Kao plant trees in future projects.*

Kao regularly protests for a variety of causes, including #SaveCongoRainforest, #ActOnLakeChad *and* #ProtectLakeVictoria, *three campaigns dedicated to African places that are suffering from the effects of the climate crisis. She also draws attention to the specific problems facing her country, such as coastal erosion, and set up an* NGO, *Green World for Better Life, as well as being part of Amnesty International Togo and* UNICEF *Africa. She also has a strong focus on global unity, bringing activists together via her Twitter campaigns for* #GlobalClimateJustice.

Most recently, she has started the campaign to #SaveTheSahel, *part of the Act on Sahel campaign, which raises awareness of yet another suffering part of Africa, and she is also working on the Africa Optimism campaign.*

With determination like this, there is one thing I know for certain. And that is this. Just as she dreamed she would, reading an article about her heroine, eleven years ago... Kaossara Sani will change the world for the better.

TWITTER: @KaoHua3

Climate change, the expression of our century. Even children talk about it. Temperatures continue to rise each day. When it's cold, we shiver even more, and when it's hot, we want to press a button to cool the Earth. Unfortunately, the famous magic wand does not exist. We must act, and stop our own massacre and suicide. Now, in Togo, things are no longer as before. The heat only increases, the dryness is prolonged, the rainy seasons arrive late, with flood damage. Populations living in the south, on the coast, are threatened by coastal erosion. Those in the north, by desertification. Farmers, ranchers, fishermen and the poorest are the ones who pay the price of global warming in Togo the most.

In Africa, like everywhere in the whole world, it is high time to take action. The climate crisis is a crisis which concerns and affects all the inhabitants of the Earth. This is why Africa must no longer be forgotten or discarded. We all have to face this global crisis, without distinction. We must act now, for Africa too. By 2100, if nothing is done, Africa will not only count the most devastating disasters in the history of mankind, but also the heaviest loss in human life. In Africa, we no longer hope to fight for our future, we fight for our present and to survive. Our waters and our forests are disappearing – Lake Chad, Lake Victoria, the Niger Delta, the Nile Delta and the Congo rainforest.

Climate change is not only accelerating poverty in all of Africa, including my country, but it is also taking the lives and hope of many people. It is stealing the future of children, who have to leave schools and become street beggars or waste collectors. Climate change affects the most vulnerable in my country, fishermen and farmers. The fishing is no longer good, the prolonged drought or floods destroy the crops, and due to lack of financial means because of the losses of the yields, families are not able to meet the needs of their children, so they stop their studies and send them into towns, where they look for work or beg in the streets to survive. Climate change is real, and it needs real action, not only talking.

To world leaders and general climate destroyers, I will only say, 'Life is more precious than money.' They don't have the right to play with our lives, just because of short-term profits. They have to take their responsibilities, and act for the planet, in the same way they would act in front of their fortunes burning in the middle of a large pyre.

About the climate crisis, I am feeling guilty, scared, angry, and that I am a victim. Guilty because the climate crisis is a collective issue, and we are all guilty of the massive murder of life and humanity. Just for our egos, we are destroying nature and killing innocent species. I am feeling guilty and angry for the fauna, the flora, for the weak and poor populations and Indigenous people all around the world. And also for everything and everyone who can't speak up about this issue. Feeling scared, and a victim, because of our modern civilization's lifestyle. We don't care about the future. We only care about money, popularity. We are selfish and think that owning material things will bring us importance in our societies. I am scared and angry because I am dreaming of becoming a mother, as a woman. But I do not want to give birth in a world engaged in self-destruction. I am scared for my next generations. Birth rates are falling more and more, because nobody wants to have children to live in a cruel world like ours. And if this continues, the human species will be doomed to extinction. Without motherhood, there will be no more children and no more humanity. Except if we want to live in the 'brave new world' of Aldous Huxley.

To all climate activists around the world, I will just say keep fighting, because change is coming and we are the change the world needs today. We should never give up, or let anyone silence our voice. We are the voice of nature, of life, of the present and the future. We are the voice of everything, and every miserable person, who is affected by climate change and cannot talk. We have a mission to stop massive murder. We are the guardians of the hearth.

KRISTINE MARIE SABATE, 19

Philippines

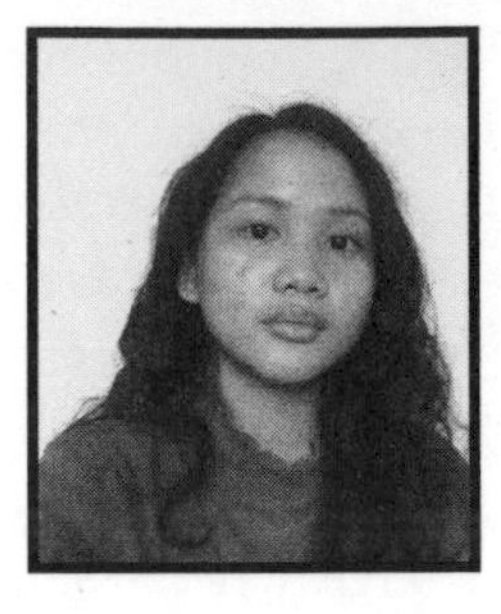

When Typhoon Haiyan (also called Yolanda) hit her area in 2013, Kristine Marie Sabate (born 2001) was not prepared. The storm destroyed her home and her community, tearing roofs off houses, uprooting trees and killing 6,000 people in the city of Tacloban, close to Capoocan, on the island of Leyte, where Kristine lived.

So, as time went on, Kristine, living in Davao City, on the island of Mindanao, became a climate activist. It wasn't a coincidence. The typhoon is a major reason for her advocacy.

In the Philippines, it isn't easy to stand up for the climate, but this hasn't stopped Kristine. She has attended the large global strikes, and joined local youth environmental summits, working with organizations within the Sustainable Davao Movement, and is associated with the Students' Environmental Alliance of Davao. She also speaks up online, sharing articles and spreading information about the climate crisis. At her school, she was the president of the Kahuy Club, their environmental group.

Kristine lives in a place which has contributed very little to carbon emissions and environmental destruction, yet is feeling the impacts here, now, and seriously. However, most of her country's citizens seem to have accepted these disasters as a part of life, because the media doesn't talk about why this is happening. So it is left to activists like Kristine to spread the word. This task is in capable hands, but it still shouldn't be up to young people.

One of Kristine's strike signs read 'NO ONE IS ACTING LIKE WE'RE IN A CRISIS'. And it's true. Almost no one is. That has to change. And it has to change now.

This IS a crisis. Typhoons are leaving communities without electricity, food or transport. This cannot be ignored. And this is only one type of crisis. One disaster. One country. What happened to Kristine is no longer unusual. These crises can no longer be called 'natural'. They aren't natural. They were caused by humans. And countries like the Philippines, people like Kristine, are the ones getting the worst of it. The ones suffering. The ones whose voices need to be heard.

So hear Kristine's message. She has experienced these disasters first-hand. She knows that this is not a hoax. It's not a joke. Act like we're in a crisis. Because we are.

Hello. I am Kristine Marie Sabate, a climate activist from Davao City, Philippines. I started my activism in April 2019. Although I do not participate in weekly strikes (as protesting in my country is very stigmatized, and you are automatically tagged as a communist or a rebel), I strike on major events, participate in round-table discussions on local youth environmental summits with environmental organizations within the Sustainable Davao Movement, and create online discussions through sharing environmental and climate emergency articles.

I am a child of Typhoons Ondoy and Yolanda (Haiyan). Being a climate activist and requesting climate action, in a country where we barely create carbon emissions yet are receiving the blow of the impact of fossil fuels, is very challenging. I want to share my experience during Typhoon Haiyan and how it created a story about my challenges as a climate activist in the Philippines.

So, let me start it off with... It was a cold dawn. No blanket was enough to provide comfort for my shivers; but the smell of early-morning Barako and toasted rice, and the monotonous murmuring of current national news, kept me warm. Most of the lights were out in the dining room, yet one remained. The news murmured some more. It caught my full attention. It was a man, in front of the green screen of a weather forecast. Unlike any other time in his segment, his tone was different. He told us to pack as many essentials as we could. 'This time,' he said, 'the typhoon is unlike any other before.' As I watched, sceptical of the message, thinking it would be one of the regular typhoons we experienced, I remembered reading the ticker tape: it was 7 November 2013 – the literal calm before the storm.

I knew something disastrous was going to happen. I just did not know to what extent. As the day of the typhoon came, the storm was indeed unlike anything before. The winds were more than I could imagine. What once was a hill covered with coconuts now became a bald, brown field. The winds threw the houses, made out of nipa and coconut lumber, to the sky, with no remorse. Roofs made out of aluminium were not enough to keep everyone dry. They flew to the sky as well. My house now became a soggy, wet shelter. Everything was drenched, even the ceilings. The wind's restlessness remained for hours. We were wet; we were tired.

Later on, the skies had already calmed. The typhoon was over. As we got out of our shelters, our emotions were all simultaneously provoked. We saw the trees in the garden all collapsed, every single one but the papaya plant; the houses in the neighbourhood were all missing their roofs; and the 7 a.m. sun had wreaked its radiance, as if it was already noon. The heat was unbearable. The damage in my village and my region was too much – so much that money was not even valuable any more at that moment, there was no food available, transportation was impossible to access, and electricity was not available for the next three to four months. Six thousand people from Tacloban, a city two hours away from where I live, died in that typhoon. It is said that the death toll was so high that some are still unidentified to this day.

Four years after the disaster, I moved to the city, and became one of the lucky ones who was able to escape the regular typhoons in the province, and to have a better and more stable lifestyle. I was able to forget the tragedy and carry on with my everyday life, as if it had never occurred in the first place.

I never knew I was privileged until now. I only just came to realize that not everyone was as fortunate as I was. While my family and I were able to receive enough to eat during Typhoon Haiyan, some of our neighbours, and other people from the affected region, were starving, injured and mourning. While we were able to contact our friends from other places and reassure them of our survival, others were devastated by the loss of loved ones and hard-earned properties.

That experience made me realize that I was taught that resilience was what we should embody the most. And that everything has a reason. I was told not to question the phenomena around me, beyond what was and is socially acceptable. I was expected not to be involved in politics, or even question current policies. After some years, I learned to cope with the trauma and not question the disaster at all. It was something that was all a memory, a natural disaster caused by the laws of nature and nature alone.

Our population is scattered over an archipelago of 7,641 islands, and we have many ways of talking, greeting, eating; yet we still have something in common – resilience. You would think at first, 'Yes, we do need resilience.' Resilience gives one hope for tomorrow. It gives the reassurance that something better will come. It gives light in times when one does not have the strength to pull through hardships; but not every kind of resilience is the same.

In this case, the Filipinos, during disasters, are too resilient. So much so, that we tend to look on a situation too positively and disregard the case for justice and accountability. But who can blame us? When systems fail, we have nothing and no one to rely on but ourselves. When no media coverage of the accountability of the fossil fuel industry, animal agriculture and other factors is shown to the people, you would not blame us for believing that no one is at fault. We will tend to believe that there is no urgency.

In my society, environmental issues and policies still belong to the 'preservation of Mother Earth' – it must be more than that. This little blue planet will carry on and exist for millions of years to come. The Earth is not in need of rescue; we are. We have to think of making the world a better place as more than hype and trendy marketing. That is something I want to help change. I want to make the conversation less of a taboo by educating (not humiliating) people uninformed about the climate crisis. I want to fill in the gap of climate education, and make information more accessible to people who are more drawn to TikTok, Instagram, K-dramas and video games. Who knows, maybe in a week or two, climate urgency is something we will talk with our parents about at the dinner table, or during a day at the park; maybe a cup of tea will create the next climate revolution.

LANCE LAU, 10

Hong Kong

Lance (born 2009) is a climate striker from Hong Kong. He strikes every week outside his school, and has been doing so since the global climate strike on 20 September 2019. He solo strikes because it often isn't safe to do large-scale protests in Hong Kong. He is also part of Extinction Rebellion (XR), and is an ambassador for vegetarian group Green Monday. He organizes monthly beach clean-ups in his neighbourhood.

Lance has started a petition to district councils, telling them to add the climate crisis to their regular working committee. In March 2020, he also launched the Green Letter campaign, encouraging conversations between property managers, owners' development committees and residents about the climate issue.

He has given talks at climate strikes, at his school and at sustainability conferences. He has also made a film for XR Hong Kong, and accepted interviews with various media to make his voice heard.

After joining a 'green visit' with legislators and councillors in Taipei, the capital of Taiwan, Lance decided to try to apply what he learned there to Hong Kong. In 2020, he aimed to build a green community, working with Greenpeace to raise awareness through videos.

For this project, Lance decided to think outside the box and write a powerful short story. It may be fictional, but it certainly reflects the scale of this crisis and the need for action. In his character, Gracie, Lance depicts a privileged person in a closeted community, who doesn't know anything about the sufferings of the outside world, until she sees them for herself and realizes that change is needed.

The story isn't meant to be totally scientifically correct. It's imaginative, but rings true. And there's nothing wrong with imagination. On the contrary. It will take thinking outside the box, imagining a better world and defying 'business as usual', like Lance has done, to build a better world.

INSTAGRAM: @fridaysforfuturehongkong / TWITTER: @fffhkLanceLau

A World Without Water

Dedicated to Greta Thunberg, Grace 'Elm'
and all the climate activists around the world.

My name is Gracie, and I am sixteen years old.

I grew up in a city called Aguas Abundantes, which means 'water in abundance' in Spanish. There were fountains everywhere, and lots of swimming pools and spas. Everyone took perfume baths at least twice a day, to make themselves smell good. People kept buying new clothes that they would wear under three times a month. We were also ferocious meat eaters. It would not be considered a proper meal for us unless there were three kinds of meat on the table.

I always loved exploring, but Aguas Abundantes had a gigantic wall surrounding it to keep outsiders out, so I could not go out to see the world unless we flew to other walled cities like mine. One day I was exploring near the wall in the rain, and then, by chance, lightning struck the wall! Bricks flew out in all directions, and a big hole appeared in the wall. My exploring instincts took over, and I slipped out of the city through the hole.

When I laid my eyes outside the wall for the first time, I saw a wasteland. It was a barren and windswept land. There was not one sign of life outside. There was a dry riverbed running along the side of the wall, leading to a murky brown sea. I couldn't believe my eyes. I thought, how were we living in such a nice city when outsiders were living in such bad conditions? I started to wonder if the wall was to keep us from the truth, too.

I trekked for a few hours, until I found a small collection of houses. I met the people there, and they said that they conserve every drop of water in the tank over there. It was all recycled water from their body fluids, and after any activities that require water, such as washing or cleaning, all the water would go back to the tank. They also told me that there were no more sea animals in the sea, because all the garment factories kept polluting the seas with chemicals used for dyeing clothes. All the clean water sources from the mountains were guarded by the security force of the companies of the big cities. The clean water flowed directly into reinforced pipes which led into the cities, to grow food for farmed animals on a very large scale. What was left was used to fill the fountains and spas.

Afterwards, I tried to get back into my city but failed. The guards did not let me in, so, with the help of outsiders, I trekked to my cousin's city for help. But then I discovered that his city had already run out of water. Even though they were still living in the city, they had to conserve every drop of it, just like the outsiders I met.

Thanks to them, I could publish this essay in the book which you are now reading, presumably in one of those big cities that are still living in a wasteful lifestyle. Now you have heard the truth. Your current lifestyle is unsustainable, and, if you continue to live this way, your city will eventually run out of water, like the rest of the world. Wake up to the truth and change now. Start with yourself. Take showers and not baths, eat less meat, and don't buy new clothes unnecessarily. Tell your friends and family about what I saw, and inspire changes.

Gracie

LILLY PLATT, 12

Netherlands / England

An English activist living in the Netherlands, Lilly Platt (born 2008) was just twelve when she wrote this essay, but she'd already had a massive impact. When out walking with her grandfather one day, Lilly noticed the huge amount of plastic in the streets. So she did some research, and found out how plastic was negatively affecting animals, the oceans and life in general. It was then, at the age of only seven, that Lilly founded Lilly's Plastic Pickup, cleaning the streets and raising awareness of plastic pollution. Three years later, she watched a speech by Greta Thunberg and decided to join the youth strike movement.

Protesting weekly, despite being somewhat difficult where she lives, quickly became an important part of her life, alongside her street cleans and plastic pollution work. She also met Greta, getting the chance to discuss the impacts of the climate crisis with the soon-to-be-famous Swedish striker.

Lilly's Plastic Pickup is more than just a litter pick. It's a chance to spread the word on social media, inform others and send a message to politicians that this isn't okay. Neither is it only limited to the Netherlands. Lilly has travelled to all sorts of places to take part in cleaning beaches. All this has managed to convince some companies to be more plastic-aware and change their policies about packaging. But this determined girl isn't going to stop. Not with the anti-plastic work, and not with her climate strikes either.

Lilly has now established herself as an international environmental champion, becoming Youth Ambassador for the Plastic Pollution Coalition, YouthMundus, World Oceans Day Italy (WODI) and Earth.org and giving talks all around the world, as well as helping others out by doing things such as standing up against bullying. In 2019, she was awarded Young Activist of the Year on UN Children's Rights Day by Comité 21, named an International Eco Hero by Action for Nature and gave her first TEDx talk in November that same year. She also started the initiative Lilly's Global Cleanup Day, an international day in April aimed at stopping plastic pollution all across the globe. She has so far picked up over 120,000 pieces of plastic, keeping them out of the sea and the plastic soup.

Her goal is to inform the world, including politicians and policymakers, about the impacts of climate breakdown and plastic pollution. Lilly is one of the top 100 influencers who tackle plastic pollution in the world. And all this by the age of twelve. Lilly proves that no matter how young you are, you can make a change. So don't ever be afraid to try. Lilly wasn't, and look where she is now.

INSTAGRAM: @lillys_plastic_pickup / TWITTER: @lillyspickup

I have been an environmentalist since 2015, when I founded my initiative Lilly's Plastic Pickup. I spread awareness through my initiative about plastic pollution, and I encourage people to reduce their plastic usage and to refuse single-use plastic.

In September 2018, I saw a video of Greta speaking about the Paris Agreement, keeping to 1.5°C of global warming and lowering CO_2 emissions. I immediately thought, 'I have to support this,' and I started my first strike for the climate on 21 September 2018. It is not so easy to strike in the Netherlands, because of strict educational rules and fines. I was lucky that my school supported me, and I continued to strike every Friday for one hour.

The geography of the Netherlands is such that it is flat and reclaimed from the sea. The altitudes are mostly zero metres or below sea level. Global warming and rising sea levels will affect life here immediately. This is part of the reason I speak out about climate change.

I was invited to the European Parliament for the climate debate and I saw with my own eyes just how politicians think about climate change. In a room of 751 seats, only twenty-eight were filled. One of the best quotes I ever heard was spoken that day. European Parliament member Bas Eickhout made a speech and said, 'If the climate was a bank, it would have been saved by now.' The strange thing was, a few months later, an Italian bank nearly went bankrupt, and a German bank did everything to keep it going. Just think about it!!!! Priorities!

When I saw Greta, we had a talk about what were the worst elements of climate change – rising sea levels, deforestation, melting ice caps or global warming... She said all of them... we can't let them get to their peak of power! The peak of power is the point at which there is no going back to change anything. We have no time machine to go back and fix them, which is WHY we need to do something NOW.

Climate laws and plastic pollution laws are always announced for many years ahead, as if they are a beacon for us to look to. For me, we need to implement regulations in the here and now. When I started striking for the climate, the talk was of the IPCC report, stating we have twelve years. We are almost two years down the line and it's ten years. Michael Mann spoke about driving on the global warming highway and wanting to get off at exit 1°C, but we passed it because we were going too fast, so we are racing towards 1.5°C. I am not sure we can get off at the exit here. This is the reality.

We just need to open our eyes to what has happened with the lockdown. It's put a stop to pollution in many ways. Look up to the sky – look at the waterways or look at the air quality figures. Nature is returning. Can we return to the old ways once the health crisis is over? I don't think so...

My advice for any new activist is to start small. Know your subject, read the news and keep up to date with the facts. Read the IPCC report. Knowledge will help you, especially when people ask you questions as to why you strike. It's important to have support – I hope it is your family firstly – and then look for your local groups. Use social media as a tool, and never take no for an answer. Keep stating your point, don't listen to people who say it won't do anything. Jane Goodall told me to never stop, always go forward, as it takes just one person to make a change. If you need any advice just message me: Lillysplasticpickup@gmail.com.

Anyone who stands up for the environment, wildlife and the planet is a hero for the planet to me. Just keep going!

LUCILA HUILEN AUZZA, 20

Argentina

Lucila Huilen Auzza (born 1999), usually called Huilen, is a climate striker with Fridays For Future, and an ambassador for the Zero Hour Movement, a protest group founded in 2017 by an activist called Jamie Margolin. Huilen lives in Jujuy, Argentina, and is of Indigenous descent, part of the Kolla (also spelled Qulla and Colla) people.

Since she was young, Huilen has been interested in nature and used to watch a lot of documentaries about various topics, including animals, space, science and the climate crisis, meaning that she has known about the state of the environment for a long time, but, she says, 'It was only three years ago that I understood that if adults were not doing anything to try to save the world, then I needed to do it. And so my activism journey began.'

Now studying legal translation at the National University of Lanús, Huilen isn't going to give up activism. Because climate justice isn't going to wait.

Her country has experienced the climate crisis a lot. Huilen says that she is lucky to be living in the capital of Jujuy, where the effects aren't as bad as in other areas of her province, but overall, Argentina's people, especially Indigenous communities, are on the front lines of the climate breakdown. The problems are massive and numerous. And Huilen isn't going to stop until life is better for everyone. Lives are being destroyed. In Argentina, in Latin America, and across the world. Activists like Huilen are making sure that those lives, those stories, are not forgotten.

Places like Argentina experience the crisis a lot more than people like me do, here in the UK. Therefore I think it is really important that we in the UK, and in all the other more privileged countries, listen to the voices of those from different demographics and poorer countries, especially Indigenous people or those of Indigenous descent, such as Huilen, because they have so much to teach us. About the climate crisis, about life, and about how to live within planetary boundaries and in harmony with nature.

INSTAGRAM: @just.huilen / TWITTER: @huiliauzza

The Effects of the Climate Crisis in Argentina

I remember telling my little sister, during our childhood, that we would never be adults, because drinkable water was going to be so hard to find that we were going to die young. So we started to take care of our consumption of water. We were children. And it did not make it easier that around that time, there was a lot of demonstration and protest against the massive lithium extraction that was happening in the Puna, along with other activities related to mining in my home province, Jujuy.

That was a turning point in my childhood. From that moment on, I somehow got to know more about the environmental breakdown in Argentina. Every day I used to learn a new word: acid rain, landslides, poisoned water, the extinction of endangered species of plants and animals, wildfires and floods during the summer. All that just in my little, remote province.

I live in the capital city in the valley of Jujuy, so I'm very lucky and privileged enough to only have experienced a few disasters just in my province. But the situation in the Puna is deplorable. The people that suffer the most are Indigenous people. They are ignored by the government, and discriminated against by the locals. What is worse is that government officials help the mine owners to try and seduce the different communities with promises of work and gifts which actually are very few, and end up destroying their lands. This is one of the main reasons why environmental refugees exist in my country.

The picture is bleak in some provinces. What the mainstream media tries to hide is overwhelming: entire villages underwater, hundreds of thousands of people with their homes destroyed, segregated Indigenous communities without access to clean water and food and neglected by the government, thousands of hectares, supposedly protected by the government, deforested to plant soy or for cattle raising. This is a true disaster, and no major news broadcaster wants or has the courage to speak up about it.

The future is worse, if what the media disseminates and magnifies the most are the losses suffered by 'the producers', who, in their vast majority, lobby to request subsidies and assistance from the government, which will surely provide them with funds. These corporations will continue to cut down the remaining forests, and try to plant soybeans even on the side cracks.

'The heart of the problem,' Pedro Peretti, farmer and writer from Santa Fe, stated, 'is the "soy-above-everything" model of production, which focuses on land exploitation and income,' a model that destroys livestock fields and natural forests. That's why we can say that the real responsibility for the floods lies with the agrarian oligarchy and its new bosses: the transnational companies that are also taking over the land. A report by the FAO (Food and Agriculture Organization of the United Nations), cited by Greenpeace on its official website, with the title 'The floods that Argentina suffers are due to deforestation and climate change', placed our country among the ten countries that have been dismantled the most in the last twenty-five years.

In response to the pressure from different environmental groups to stop the massive deforestation, Congress passed a law in 2009 known as 'Ley de Bosques', which divided the country's forestland into three sections: red (untouchable), yellow (mixed use) and green (available to deforest). But the deforestation continues, daily and in many provinces. Throughout 2017, 42 per cent of the deforestation in the country took place where regulations did not allow it.

Unfortunately for some communities in Salta and Chaco, the different local governments have unfairly authorized vast deforestation projects in red and yellow protected zones. In Salta, for example, more than 170,000 protected hectares have gone under the bulldozer since the law came into force. The death of Wichí children in the province of Salta, the lack of recognition and titling of their territories, deforestation, lack of access to safe water and poor food assistance are the main causes of climate refugees, and malnutrition, causing deaths. It is estimated that in the Gran Chaco region there are around 200,000 Indigenous people from nine Indigenous tribes, mostly Wichí and Qom.

People only look for better living conditions, so when they migrate, they do not attribute it to the climate, they say that it is because there is no work, because the land no longer performs as before; but the primary reason is the climate crisis.

Malnutrition as a consequence of food shortages, along with decreased quality and quantity of water and the deterioration of the environment, are and will be conditioning factors that will contribute to the appearance or worsening of diarrhoea, a leading cause of death. Abrupt climate change also increases pulmonary and respiratory diseases. All these conditions influence morbidity and mortality, above all among the poorest boys and girls.

Forced migration due to climate change is not the same from one culture to another. These have their particularities due to their causes, the diverse ecological and human contexts that deserve their own analysis, diagnosis and treatment. There is no single answer to these phenomena.

Migration from poor countries of the Global South, such as those of Latin America, Asia and the Asian subcontinent, towards the rich countries of the northern hemisphere and the emerging ones, will continue; although south–south migration and migration across borders have increased.

The truth is that the drama is not only local, and won't be understood if it is not linked to the concentration and mistreatment of the land by transnational companies and the absence of a strong state government, all throughout the history of Latin America. Poor countries, like ours, are the most affected, and although we cannot face the problem in isolation, because it is a global phenomenon, international solidarity seems to be lacking. Rich or emerging countries are better prepared to mitigate climate impacts.

The only answer to control the various impacts of climate change is a global agreement among the largest greenhouse gas emitters, for which rich countries are not prepared, because they state that it would prevent the economic growth that they want so much to balance their economies, particularly now that they are drowning again in financial crises – of employment, economic growth and a global recession. Consequently, monitoring the progress of climate change is becoming a second or third priority.

MADDY ANNA BURTON, 17

England

Maddy (born 2002) is a youth striker and YouTuber from Manchester, England, UK, who has been attending monthly strikes since early 2019. A talented photographer, she is mainly involved with taking pictures at the strikes, and some of her photos have been shared by her local strike groups on social media. She also does some organizing / planning within the Manchester youth strike group.

She was inspired to start striking by Greta Thunberg's weekly sit-ins outside the Swedish Parliament. She said it was the first time she had seen someone she could 'relate to about my worries towards climate change, as she was the same age as me and felt angry and helpless about what to do'. This shows how Greta's age has made climate activism much closer and more relatable to young people, and showed us that we can change things, too.

Maddy knows that, living in the United Kingdom, she is very privileged when it comes to the climate crisis. But even so, the effects are being felt. Maddy has witnessed flooding, which destroyed infrastructure, displaced people and claimed several lives, showing that climate breakdown, in her words, 'really is a global emergency', even though it is, of course, affecting some places far more than others.

For Maddy, the biggest driving motive for her activism is the issue of equity / climate justice, which means that richer countries need to reduce emissions much faster, to allow poorer countries to develop their standard of living, and is, more generally, about the aspect of equality, which is 100 per cent tied up with the climate crisis. Without equality, then, even if we 'fix' the climate crisis, we will just keep making the same mistakes that caused global heating in the first place. Gender equality, racial equality, equality for all, is vital in this fight. There is no true climate action without social justice.

Maddy isn't 'high-profile' or famous, but she is a vital part of the generation of change that is fighting for a better, brighter world.

INSTAGRAM: @maddy_anna_ / TWITTER: @maddy_anna_

I have been going to the monthly strikes in Manchester since February 2019, and ever since, the climate crisis has always been in the back of my mind. I think people often don't realize how much genuine fear I have for the future of our planet, my children and grandchildren, the hundreds of species going extinct every single day. This is the biggest issue we face, not only as a country, but as a planet, as humans, and it frightens me that nobody seems to be acting with urgency. Why is this not the top priority of every politician, in every government, in every country? Why is everybody not constantly talking about it?

As someone living in the UK, one of the richest countries in the world, you would perhaps expect that I couldn't possibly be affected. We're generally a very lucky country, in comparison with so many others. And this is the argument people often use when trying to ignore the climate crisis: 'I can't see it, so it can't be true.' Yet in the summer of 2019 Whaley Bridge, a village very close to my home town, was hit by a flood, causing the dam to collapse and forcing people to evacuate their homes. Following this, the floods of November 2019 which happened all over the country shed light on the fact that this really is a global emergency. Having said that, our plea for climate justice draws on the fact that, the majority of the time, it is the countries who contribute the least to climate change that are the most affected by it.

We need a worldwide effort to push the issue to the top of everyone's agenda. We need system change, starting from the top, with the people in power to implement it. So that's why, when Boris Johnson says the floods are 'not... a national emergency' and doesn't turn up to the climate debate, and when Donald Trump withdraws from the Paris Agreement, I begin to lose hope. How can we ever tackle climate change when some of the most influential people don't care? How do we get our voices heard above the constant shouts of Brexit, or the capitalist-driven campaigns that put their aims of short-term profit over the long-term needs of the planet?

I read recently that just one hundred companies are responsible for 71 per cent of global emissions, and this just seems ridiculous to me. That is such a huge number that, as individuals, we can't really do much about it. Yes, individual change does help in small proportions, and we should all be trying to cut down on our own carbon footprint, but surely this shows that the root of the problem lies within the already extremely rich companies, who seemingly only care about profit? Therefore, we must then turn to our flawed, capitalist way of life. We need a full shift in the way we see society and a reshuffle of the things we care about the most. It is no longer an option to prioritize meaningless sales and the greed of consumerism over saving our animals, preventing extreme weather and sea levels rising, and protecting the more vulnerable people who cannot buy their way out of these crises.

Being a part of the youth strike movement is something which I will never take for granted. To see so many people come together every month who share the same fears as me truly makes me hopeful. I feel so empowered to have a voice in demanding climate action, along with so many others, and it does feel as though we are making progress. Maybe not as fast as we'd like, but it is there. It is incredible to see how much the movement has grown since my first strike in February 2019, and I have seen changes in the way people see and talk about the climate emergency, whether that is on social media, in the news, or just in the conversations I have had with people. But it can't stop there.

My message to world leaders would be that, whatever your political views may be, and however rich and privileged you are, we need you. As youth strikers, we've done our bit and brought the crisis to everyone's attention, at least those who are willing to listen. We can give speeches and march through the streets and shout as much as we want, and we will carry on, but without powerful leaders who back us, there's only so much change we can make. And yes, it might seem like a huge barrier to overcome, and it is, but if anything is actually going to change, we need radical decisions. Waiting for my generation to fix your mess is no longer an option. We have ten years to cut down on our emissions before it's too late. Ten years is not a lot of time, but it is still possible. Difficult, yes, but possible.

Ten years after writing this, I will be twenty-seven. I might even have children of my own. And if they are forced to grow up in a world that we have destroyed, despite having the power to save it, I will never forgive us.

MAUREEN DAMEN, 16

Senegal

Maureen Damen (born 2003) is a teenage climate activist from Dakar, Senegal. She is one of the co-founders of FFF Senegal, the Senegalese Fridays For Future group, with fellow activist Yero Sarr. Maureen also founded the Senegalese branch of the Rise Up Movement, her country's chapter of the movement founded by Vanessa Nakate (whose essay appears later in this book), which centres around empowering African climate activists, who often struggle to be heard in this fight, which tends to seem more white and European.

Every day, Maureen strikes for the Congo rainforest, raising awareness of the deforestation and clearance that is destroying this lung of the planet. But she isn't just doing this herself. She's inspiring others. Helping friends to join is super important and is the reason why many people feel able to protest, so this is just yet another reason why Maureen is totally amazing!

As well as daily Congo strikes and having helped found two groups, Maureen strikes every week for Fridays For Future, holding signs that are simple but effective, with messages like 'Sauver La Planète' (Save The Planet) and 'Act On Science', as well as catching slogans like 'Poll_tion, Ed_cation, Sol_tion. What's Missing?' (Hint. It's you!) She also speaks up on Twitter about other issues facing Africa, such as the shrinking of Lake Chad and pollution in her home country.

As if that wasn't awesome enough, Maureen is also a human rights defender. She also encourages people to look after their mental health and not be afraid to take breaks.

Through her work with Rise Senegal, as well as her independent activism, Maureen is helping more Africans join the climate movement. Making sure that Africa is represented is so important to her. She says, 'We need more Africans in the climate movement!'

Senegal is already suffering from the climate crisis, and specifically from pollution due to inadequate waste disposal. In her city, Dakar, Maureen sometimes has to stay inside because of the terrible smoke-filled air. She knows this isn't good enough, so she's choosing to do something about it. You can do something, too. Whoever you are, learn from Maureen's story, and, just like her, start to act.

INSTAGRAM: @damenmaureen15 / TWITTER: @damenmaureen15

Climate change is affecting Senegal enormously, in different ways. For example, we have a landfill dump called Mbeubeuss that contributes to the pollution of Dakar. Although it is all surrounded by inhabitants who are aware of the dangers of this defective discharge, it no longer works very well and constantly pours the waste out around it, or else spills an incredible black smoke, visible to the naked eye several kilometres away. We don't own good landfill disposal facilities, so people here are used to burning their waste outside, sometimes in public areas.

In my school, we have a lot of trash, but the students keep on throwing everything on the ground, and I believe it's all about education. Adults, through generations, haven't taught their children to throw their waste in the bin, that's why today everyone is throwing everything on the ground, without minding that they are destroying the ecosystem. You can see some streets in Dakar where it's totally clean, because the well-to-do people live there. There is trash disposal on the street corners, there are multiple alternatives on those streets, but not in deprived neighbourhoods. But in the deprived neighbourhoods, there's a lot of pollution, and it's sad and poorly maintained.

My message to climate destroyers and leaders would be this:

For years, you have been debating the same subject. For years, you've been finding solutions without even applying them, destroying the planet day by day. The solutions are indeed in front of you, but you refuse to bend to them. Unfortunately, the climate destroyers do not seem aware of what they are doing. Only you can change the world. You can no longer continue to ignore the facts and destroy our planet, which is so fragile, as if nothing had been done. But it is not too late to act, right now we can set the record straight. Listen to scientists, and adopt the necessary strategies to preserve our environment sustainably. This is not our planet definitively; we are only borrowing it from future generations. Its fate is now in our hands.

My feelings about the crisis:

The climate crisis is unfortunately a crisis that affects the whole world, whether we like it or not. A crisis that is not taken into account by a large number of people and is even neglected. I find it completely childish, but it is understandable. Understandable from the point of view where not everyone is aware of this crisis. Indeed, it is little discussed. It is for this reason that even before, I did not know much about the environment and all that surrounds it. I was not paying attention to it, because I had never been brought up to be aware. This is a serious mistake that creates a lot of careless attitudes about it. If the government proposed environmental programmes in school instruction, more people would know the subject, more people would want to act, etc. The climate crisis is a crisis affecting generations, a crisis that should have been resolved a long time ago.

My advice:

I encourage young and old to join the climate movement. Not to listen to negative comments and hatred. To contribute to planetary justice, to raise awareness around oneself. Not to hesitate to take time for oneself, to rest. To be confident and to go to the end, even if sometimes it seems difficult. To listen to, and share, what scientists say about climate change. We're going to have this global climate justice. We are in this together.

MEG WATTS, 20

England

Meg Watts (born 1999) has been a part of many different environmental activist groups over the years, namely Greenpeace Bath, XR Youth, Bath Youth Climate Alliance and now Radical Restart. She aims to emphasize the 'intersections of global politics and environmentalism', through her work on decolonizing environmentalism and the importance of global climate justice. She was also the president of the University of East Anglia's Sustainability Society, planning to rewild the campus, creating a discussion space for decolonial environmentalism and connecting with other activist groups and resistance struggles internationally. She has run various campaigns raising awareness of global climate justice through writing and graphics. She also ran an anti-fast fashion photography project called All Secondhand, which 'aims to highlight the ease with which we can alter our shopping habits, promoting individual action alongside corporate and political overhaul'.*

Meg has been involved with nature and the environment since she was a kid; one of her parents is a botanist, and the other is a teacher who specializes in ecology. She grew up pond dipping, hedgehog spotting and visiting botanical gardens, meaning that she has always loved and appreciated the natural world. Watching David Attenborough documentaries and reading 'cli-fi' (climate fiction) further taught her to hold nature dear and stand up for the environment.

Meg's activism started with being the environmental secretary at her secondary school, and then ended up turning into volunteering, photography, graphic design, writing, protesting, striking and event coordination with all the numerous different groups I mentioned above.

Writing being her passion, Meg is building a portfolio of work that 'aims to help redefine our relationship with (and perception of) nature through creative exploration and accessible climate theory'.

Although the United Kingdom doesn't suffer too much from climate breakdown, Meg poignantly describes seeing the terrible drought of 2018, flying home over London.

Meg chose to stand up and take responsibility, for her personal actions and for what countries like the UK have done to the planet. She knows it shouldn't be up to her, as a young person, to take so much of this on her shoulders. But she is doing it anyway, out of love for nature, and this beautiful planet, and our global family.

INSTAGRAM: @megwattscreative / TWITTER: @megwattsmakes

Flight Shame, Climate Justice and the Activist's Handbook: Organizing Global Protest in a Climate Breakdown

It started with the bees. I'm sure there were more, at least I think there were more, back in the summers of my early childhood. Then there were the 'please switch me off' posters by the light switches at school. Then there were the dramatic ending scenes in nature documentaries, the David Attenborough showcases of diminishing forests, and species threatened by extinction. The climate breakdown and associated crises of global climate injustice have always been an intangible, looming threat within my life.

I am fortunate to live inland, in a country that has (so far) been minimally affected by climate change, in comparison to more environmentally and economically vulnerable places across the world.

My experience of climate breakdown was, up until recent years, purely observational. I got my information through the campaigns of NGOs, like Greenpeace and WWF, alongside heated TV debates over the validity of the IPCC's stark warnings. The climate breakdown was most visible to me when I was flying home over London from abroad in the summer of 2018, barely holding in my horror at the bleak scenes of drought, and lack of living green, in the city below me. I remember being paralysed with guilt in this moment, knowing that my actions were contributing to this heatwave and other irregular weather patterns like it; my two long-haul flights, my carbon emissions, were endangering the lives of millions of vulnerable people across the globe.

But my experiences of the climate breakdown are nothing in comparison to others in our global community. There are people who deserve this platform more than me. People who have seen more, who have felt the losses of climate chaos, who are on the front lines, defending the environment against exploitative industries. That is why it is of the utmost importance that we privileged few, who are yet to feel the full brunt of the climate crisis, must campaign for others across the globe. That is why we are striking, and protesting, and organizing. Because the youth know what's coming, and we see what is already happening.

If I could wrangle all of the world's key decision makers, political leaders and corporate CEOs into one room, ensuring that they would actually listen, I would give them a very simple message from the UK.

Two years of geography A level have given me a more accurate understanding of climate change and global systems than, I would hazard, most party politicians have. This is shocking; the fact that just two years of education outweighs our current leaders' understanding of climate change (and the urgency with which we have to act) demonstrates just how underprepared, misinformed and short-sighted both our government and our society are. We need change, and we need it fast. The next months are crucial in preventing positive feedback loops, loops that will cause an exponential and potentially unstoppable rise in global temperature. I would therefore rather sacrifice my luxuries now, to prevent warming above two degrees, than live in the world of catastrophic weather conditions, thinly veiled eco-fascism and mass extinction that this will bring. Whether we like it or not, this is the reality that we will have to face, and it's a lot nearer to all of us, globally, than you might think.

The younger generations are watching, and we are going to hold you to account. We are trapped in systems of power and consumerism that prevent us from living as ethically as we truly wish to, but we are changing as our new reality forces us to change. We are not interested in short-term profits. We are interested in sustainable, equitable, harmonious survival. Your greed, your denial, your short-term thinking and your inaction are responsible for so much destruction and misinformation in the pursuit of exponential consumption and growth. We youth are asking you leaders to collaborate, transnationally, to reduce global emissions. We are asking you to listen to the experts. We are asking you to devote your resources to this crisis, with the urgency befitting a crisis.

I want leaders to know that, even as they fail us with their inaction, we young people are acting as lighthouses to our peers. We will act as lighthouses to our children, if it is safe enough to have them. You have forced us into this position of responsibility. We will not succumb to your false politics of overpopulation, or your xenophobic fears of resource insecurity and consequent migration. We are ready to dedicate everything to stop your destruction of the planet, short-term educational prospects and long-term job prospects included.

As a child and young teenager, my emotional response to the climate crisis was characterized by desperation and fear. I still have that desperation. I still have the sinking feeling that despite all my efforts, and the efforts of other activists, those in power will refuse to listen to us until after it is too late. However, thanks to the mobilization of millions of young activists, I now have a new kind of hope. A stalwart determination. I know I will do the best that I possibly can to prevent warming above two degrees, with the tools and the voice that I have. I am fully dedicated to that end goal. I am devoting my life to communicating the realities of the climate crisis. All the work that I do, I do for the planet I call home, and the people I call my neighbours.

I would encourage anyone who feels similarly, whether in terms of fear for the future, or in terms of determination, to act on their impulses and start campaigning for global climate justice and equity. There are many levels to this commitment. Personally, in terms of your individual action; socially, in terms of your close peer group and family; and globally, in terms of the international implications and optics of your actions.

For personal environmentalism I would encourage any aspiring activists to try as best they can to align their actions with their values. In my experience, if you know you are doing the best you possibly can for the environment, in your situation, it's far easier to trust your own opinion and ward off the creeping combination of eco-anxiety and guilt. It can be simple: eat vegan where possible, carbon offset, buy second-hand, avoid car journeys where possible. These changes are often difficult for others to adapt to, and for us to maintain; we live in a system that makes ethical living inaccessible and expensive, and consequently the most ethical choice is not always viable. This is something we can fight against slowly, through our function as role models to our peers. It is also something we can campaign for on a political and consumerist level, supporting green politicians and boycotting unethical companies.

It is important to remember to use your voice to object to (and reject) systemic environmental exploitation and inequality, even when you are unable to fully escape the system. I would encourage you to educate yourself on issues of ecology, carbon science and climate equity, from a variety of sources. A diverse and multifaceted frame of reference will allow you to accurately explain both ecological and political issues of climate change to others, with confidence and credibility.

Finally, in terms of participating in youth movements or orchestrating your own actions, I would encourage you to collaborate and collectivize. This way the load of grassroots organizing can be shared and the optics of your planned actions considered from a variety of perspectives. I always have five main questions when planning or evaluating the success of an action:

1) Is this activism effective? Are you getting the scientific message across as clearly as possible while maximizing impact and minimizing carbon emissions?
2) Is this activism researched? Are your facts and statistics accurate? Would you be able to support and defend the necessity and urgency of the climate movement through this action?
3) Is this activism accessible and inclusive? Is there space in your movement for people of various backgrounds, ages, languages, levels of mobility and financial situations? Is everyone accounted for?

4) Are you considering global voices? Is your activism purely focused on the struggle of the future, ignoring the experiences of climate activists and refugees right now? A global problem cannot be combated without global awareness, global solidarity and global cooperation.
5) Are you considering the implications of your activism? Are there obvious elements that could be negatively portrayed or misconstrued by media outlets? We as activists are trying to promote the climate movement, not distance people from it, therefore we must consider the optics of our actions.

This is a crisis we are facing together, as a global community. This is a crisis that can only be solved through global collaboration, global investment in environmental science, and a global will to live with long-term environmental sustainability in mind. We youth are asking you to think beyond your personal wants and needs. We are asking you to see the bigger picture. We are asking you to join us. We need you, and we need everyone else on the planet, to campaign for global climate justice.

MICHAEL BÄCKLUND, 17

Israel

Climate activist Michael Bäcklund (born 2002) risks his future prospects to fight against the climate crisis. Climate activism for youth is difficult around the world, but it is particularly challenging in Michael's country as there is a lot stacked against him. He says, 'Ranging from a strict school system to social norms and political instability, climate activists in Israel can be easily brushed aside.' However, this hasn't stopped him from standing up for the planet.

Part of ClimateScience and Strike4Future Israel, which is the branch of Fridays For Future he works with, Michael organizes climate strikes and works globally to standardize reliable climate education in schools, institutions and society. His country, being in the Middle East, which is, as Michael says, 'a climate hotspot', is already seeing the effects of the climate crisis. Massive floods are destroying people's lives every year. And Indigenous people, living in the desert, are being severely affected.

But despite all this, his area is not very climate-aware. Especially when it comes to politicians and elections. Partly due to the ongoing conflict, none of them seem to care about the climate crisis or the future of the youth. This is what Michael is up against, and he is trying his best to raise awareness and pressure them to act. But it isn't easy. He has been ignored, brushed aside and told that it is up to him and his fellow youth to save the world, when in reality there isn't enough time for our generation to be the ones in power. The world can't wait for us to grow up. That doesn't mean we aren't changing things, however. Raising awareness, protesting, will pressure change. But the longer the adults shift the responsibility onto their children, the longer action is delayed and we sit around listening to excuses.

Although Michael's part of the world is already suffering from the climate crisis, things are only set to get worse, with extreme heat becoming the norm in an overcrowded country, which will lead to all sorts of catastrophes such as water shortages and crop failure, not to mention the fact that it will escalate the conflict, costing lives.

So adults, stop saying we young people will save the world and using it as an excuse. Instead, listen to us, and act on our words. And the good news is you can start right here, with Michael Bäcklund.

TWITTER: @Mina_mihku

Hey everyone. My name is Michael Bäcklund, and I am a climate activist in Israel.

I feel like I cannot live a normal 'teenagehood' under the existential threat that's called the climate crisis. Many times, I have found it hard to sleep because I stay up late, thinking about its consequences. I think about our movement, and if we are going to actually change something or not, or if I'm doing enough.

Then I wake up, and I continue my work, which occupies all my free time. Every night, instead of sleeping, I read governmental agreements and trade deals and scientific studies. I have read over 2,000. I have had to skip school and suffer from pressure and work stress as a sixteen-year-old without any pay, because the adults who are *getting paid* are not doing the right thing.

One thing I can't put my finger on is who would want to put, on a child, the pressure of the world, our existence and everything we call life, and, where I live, I don't have many people to share this responsibility with.

In European and other countries, climate politics are limited and not developed enough. In Israel they are non-existent. This is a result of multiple different things, such as lack of information and political instability in our country, which refers to the ongoing Israeli–Palestinian conflict. Environmental legislation is considered irrelevant by the major parties, since it doesn't win elections and it doesn't address the conflict. In the last elections on 2 March 2020, there wasn't even a single Green party involved, nor a party that had a Green New Deal. War usually polarizes both sides of the parties, and people and basic aspects of life get blurred by hatred and over-focus on an issue.

In the fourth annual climate summit of Israel, in 2019, I asked the Minister of Energy a question: has the ministry officially given up on the future of the Israeli youth, and does it disbelieve science?

The minister told me no, and that it was up to me to fix the situation, and proceeded to ignore me.

On top of that, as I have mentioned before, my whole life is now taken over by climate activism. This means that some other things suffer. This particularly shows in my school progress. In Israel, it is made extremely hard for teenagers to strike, since they will be high school dropouts if they miss more than 30 per cent of the yearly lessons, which means that if one were to strike weekly, one could only be sick for ten days in that whole year, or they would be a high school dropout.

Weekly striking and protesting has been made risky for people like me, and that's why many people feel trapped in their activism, because they want to express themselves and help in the fight against the climate crisis but fear they will be left without a high school diploma, which in Israel has a massive impact on possible careers in the future and in the army, the joining of which is mandatory.

I also happen to know that Indigenous Druze and Bedouin populations who live in the desert are heavily affected by climate change. I once had a friend who told me that they have an olive tree farm in their village, and every year for the past five years, there has been a significant drop in the harvest, which directly impacts their financial stability.

Every year, there are massive floods all over the country. A person in my school died a year ago when he was on a trip with the army, in the desert, in a flood where he tried to save other people. That day seven people died in total. In January 2020, six people died in floods that crippled the economy of a city, and no media or party linked it to the climate crisis. Even if someone doesn't die in these floods, their cars do, and this can be a big financial setback for students who don't have massive incomes and have used all their savings to buy their car.

It is estimated that, in Israel, the population will rise to 16 million by 2050. It is also estimated that by then it will be more than 50°C by day, for most of the year, because the Middle East is a climate hotspot. For some context, there will be 16 million people in a country that is 27.5 times smaller than Germany. This is such a terrible reality, which makes it actually hard to comprehend, and this, too, keeps me awake at night. These extreme circumstances will certainly escalate the Israeli–Palestinian conflict and these conditions might make my home country uninhabitable.

I would like to go to school, get a uni degree and live a steady and happy life, but apparently in the twenty-first century that is too much to ask.

Dear leaders of Israel: Please pay attention to the climate crisis, and don't forget the upcoming generation. Climate change is not just a threat to biodiversity, it also affects us, and has the potential to ruin the world and the country we live in. Unfortunately, time is ticking and I know you have not started talking about this, nor negotiating this in the Parliament halls, but there is no time for that. We need action now. Please listen to the scientists, and do the only sensible thing, which is to recognize the climate crisis and move to renewable energy as fast as possible. Build storage and support innovation, so we can make the renewables even cheaper.

To all the other climate activists of the world, I thank you for standing beside me, whether you are living abroad or in Israel. You are the reason I haven't cracked under pressure.

In case you are not a climate activist yet, you are not too late. We need you, and I need you. We are in a war against climate denial, ignorance and, in some cases, blatant corruption. But we have the greatest weapon of all, which is the truth. We have the truth, and a scientific consensus, and if we don't spread the word, I'm afraid that all we have learned to love might be lost, in just a window of ten years. We cannot allow the world to reach the point of no return.

MOLEMO BIANCA KGANTITSWE, 17

South Africa

Molemo Bianca Kgantitswe (born 2002) is a climate activist and writer from South Africa. They hadn't yet attended any protests at the time of writing this essay, but they have made changes in their personal life, as well as encouraging those close to them to do the same. They are also involved with various activist chats and online groups, and are an advocate for equality.

They were inspired to start activism by celebrities who spoke out about the issue. Although that's not many, Molemo Bianca says that 'the handful who did, made an impact on me'. They started becoming involved with the climate issue about two years before they started activism. They would have discussions in their biology class about the climate crisis, and then started reading articles and doing research on it. However, they didn't start acting until early 2019.

Another big reason for their activism is how the climate crisis has massive links to inequality and discrimination. 'Climate change not only impacts individuals universally... it impacts even [more] underprivileged masses, the people without the recourse to escape the effects... Places of high [levels of] drought and deforestation, are usually places where underprivileged people live.' An example of this is Molemo Bianca's grandparents and extended family. They would visit them, and see first-hand the effects of the climate crisis.

Their activism so far has been centred around the 'Rs' – reduce, reuse, refurbish, recycle. They did this themself, and also got their family involved.

As well as environmental issues, Molemo Bianca also advocates against sexual abuse, racism, gender-based discrimination, and the oppression of the LGBTQ+ community.

They aren't scared to speak up for the causes they care about, and for the issues that are important. Everyone should take a leaf out of their book. Stand up, stand out, speak up, and show you care.

INSTAGRAM: @iokrysie / TWITTER: @asgayasthey

It's always going to be a matter of what the environment means to me, how far my voice can carry my worry and my passion, how long I'm willing to preach the same thing – around me, online, or to strangers. However, one thing remains: it's always a matter of how I'm doing my part.

The simplest thing I've done was walking, and driving less. This aroused an interest in me of how beneficial this is, for the environment, and, selfishly, myself. Then came the use of less to no plastic, then came the reusing, the recycling, the ability to be conscious and research and take the time to be resourceful.

Reforestation is a new project I'll be pitching to my school, and recycling is the new theme I'll be taking my own time and money to instil into the school, and my parents' home before I permanently move out, so when I'm alone and reusing, there's a reassurance that they are too.

I'm almost eighteen as I write this and there's a lot I don't know, that I'm googling and reading and watching. The melting glaciers aren't a myth, but people, hearing me talk about these issues, are surprised that it's happening. It is almost as if people cannot grasp the damage they're doing, and that, despite the slim chances, it's irreversible.

I have never been to a protest or peaceful march for the environment. But I've made plans to this year. Not only that, but I was interviewed at the beginning of the year, and that boosted my story and it gave light to what I stand for and what I encourage greatly.

My activism doesn't serve to place people of colour in a position to have to voice their concerns and their point of view in the space of white people, because, in my own opinion, trying to unite people who are oppressed, with their counterparts – who benefit more than them and have unlimited spaces – is unfair.

Unity is important, but so is being able to find the underlying issues, and race seems to be one for the majority. I want unity for the concern of our environment, but we need to tread carefully in terms of who is more affected.

Are white people conscious of the crisis and the effects it has on Black marginalized groups?

I think the white population does care, but to the extent of the environment, and not necessarily those affected and how to tackle that issue for the best of them and their interests.

Are people informed? The youth is actually not as informed as older people. Granted, the youth has access to the information, and knows how and what to search for more than the older people, but it's a matter of whether enough of them care or not, and then there's still the aspect of what work they're willing to put in to know.

My country and the effects global warming has on it: South Africa, I believe, is most prone to the effects of climate change in more disadvantaged areas. People who depend on nearby rivers for water, who grow their food, who depend on weather for their farming, etc., that other people depend on, are really impacted more than we can imagine.

The Fridays For Future movement in Africa is based on our pollution, and how we choose to reuse, reduce, refurbish, etc. Recently there has been flooding, and dumping sites are almost everywhere, also food doesn't grow as it should any more. It's a call to the government to pay attention to the outcry of the environment as a whole, and those affected.

As much as we can focus on climate change as a whole, white people are still very privileged, in the sense of not necessarily experiencing what my people do. It's still an ongoing discussion and fight against the accountability and the actions they need to take.

The environment isn't just a pretty sight, it is more than an aesthetic for our phone wallpapers.

It's what we breathe, eat, live and embrace. And what if we can't take care of the one place we are able to inhabit? What does that say about us, more than it does about the place we are in?

This is a question that drives me every day to make a change. No matter how small, and hopefully as big as can be.

NASRATULLAH ELHAM, 18

Afghanistan

Growing up in war-torn Afghanistan, it took Elham (born 2002) some time to realize that climate change, as well as war, was affecting his people. But once he did, he knew he had to act. He was a local activist for a few years before hearing about Greta Thunberg and initiating Climate Strike Afghanistan, a strike movement in his own country.

As well as fighting for the climate, Elham is the founder and president of Laghman Peace Volunteers (LPV), a group fighting for a peaceful Afghanistan. He is also a feminist and human rights advocate. Basically, he always speaks up for the issues that matter.

When he went to study in Thailand, thanks to a United World College scholarship, Elham didn't stop acting. He helped Climate Strike Thailand, by being their coordinator with other organizations, and organized climate gatherings in Phuket. When his studies were over, Elham was ready to head back home, but then he got stuck in Thailand due to Covid-19. However, he continued to work, online, with the Afghan strike group.

Elham became a climate activist after seeing the loss of human lives and livelihoods in his country, which is already being heavily affected by the climate crisis.

Overall, there are so many issues facing Afghanistan, and Elham doesn't want people to ignore them any longer. He is frustrated that many privileged world leaders continue to deny the climate crisis, thus causing the deaths of more of his people.

By fighting for peace, climate justice and human rights, Elham hopes to make the future of Afghanistan, and the world, a much brighter one.

FACEBOOK: Nasratullah Elham / TWITTER: @NasraElham

When we were living in Mihtarlam city, in eastern Afghanistan, we were lucky enough to at least have schools open. My previous school in the village, about twelve kilometres from my current house, was already destroyed, and insurgent groups had turned it into their stronghold.

As unbelievable as it may seem, going to school in a turbulent atmosphere of actual bullets crossing the sky was almost a normal experience of my childhood. Sometimes, it was so dangerous that my mother would rather make us stay home than go to school.

Born and raised in one of the most war-torn sites of Afghanistan, I developed a deep understanding of the problems that the young generation of a nation at war goes through. Climate change was nowhere near an issue of my concern. Therefore, unbeknown to me, deadly cyclical floods, a significant decrease in average rainfall, an increase of 0.6°C in the average annual temperature and in the frequency of hot days and nights, as well as harsh droughts, were already foreshadowing a death sentence. One that was to come soon, as a result of climate change, for this landlocked country right in the heart of Asia.

People in my country, despite being on the front lines of the devastating impacts of climate change, happened to know little to nothing at all about it. This was due to a lack of awareness, that, in many ways, stemmed from the absence of adequate education. Though crucially, it was also indisputably important to prioritize other issues more pressing, such as finding ways to save our lives: to find a piece of food to survive or find something to bandage our injuries. And so we were unaware of the fact that climate change greatly inhibits the solutions to all these problems.

Afghanistan has been named one of the countries most vulnerable to the devastating impacts of climate change. I have heard so many elders bragging about how they were able to get much greater yields from crops in the past, because they were more hard-working than today's farmers, but, in fact, there isn't half the moisture available in the soil during the growing season today as compared with those times past. Crop failure is more frequent than ever before.

The decrease in rainfall during cultivation, and crop failure due to these water shortages, has made such conflicts far more common and most frequent in the west of the country. Many of these conflicts lead to some of the worst human rights violations, most notably child marriage and related forms of violence. This happens as farmers often believe they have no other option but to exchange their young daughters to landowners as a means of compensating for the lost income.

I have personally witnessed conflicts between landowners and farmers in my village, where some families have had to sell their household assets, often their only means of transportation, usually a motorcycle – an essential item in rural areas.

With 80 per cent of Afghans relying on agriculture for their livelihood, and 70 per cent of the country's irrigable land predominantly relying on run-off water, increased water scarcity due to fast snowmelt, severe droughts and, in a different season, soil erosion due to flash flooding, all contribute to a disastrous level of food insecurity.

And when scientists predict an increase in the number of hot days and nights, and a 4.0°C overall increase in the country's temperature by the 2060s, it scares me to think of what this will mean for a country where barely one in three citizens has access to twenty-four-hour electricity and where thousands of lives are still lost from curable conditions like diarrhoea.

Here in Afghanistan, it's not about the coral reefs dying or bush fires, and of course, this landlocked country, surrounded by tall mountains, won't be wiped out by sea-level rise, but instead, we will be dead inside.

In Afghanistan, the impact of climate change goes way beyond ecological factors. It fuels insurgency, escalates conflicts, exacerbates starvation, increases displacement, encourages child marriage and blocks our prospects for peace.

In many ways, climate change has been destructive to the Afghan cultures, civilizations and traditions that have evolved throughout the centuries. Take the river of Oxus for example, which has been the cradle of civilizations for thousands of years, or the river of Panjshir, which occupies a few stanzas in almost every poem from northern cultures, or maybe think of the Helmand river basin, where almost every legendary mythical Afghan couple has secretly dated. These rivers, or, as known by my people, the precious blessings of the divine, have turned into real and tragic monsters, taking lives by the hundreds of those residing alongside them, washing away their land, shelter and almost anything they have. Such disasters have become far too commonplace, and more cyclical than ever before.

Our white mountains of the Hindu Kush with the beautiful rays of sunshine, which load up our rivers with fresh water, are not joining us for the Nowruz celebration of the spring as they used to... instead, the too-fast snowmelt in early spring puts us at increased risk of this flash flooding and a limited water supply for the rest of the year.

Life is beautiful listening to the Uzbek, Turkmen and Hazara women singing folklore with their daf hand drums, celebrating the ancient tradition of carpet weaving. But it brings great sadness to me to say that this is not any more, either. The mass starvation of animals, a completely disastrous loss of livestock due to severe droughts, has led to a shortage of wool, and their artistic fingers are not able to weave any carpets, which, besides being an integral part of the culture, is also a primary source of income for many families. Afghan carpets have received prestigious recognition in world markets and are among the most expensive in the world.

If it's spring, thankfully it does still mean that our vast deserts are dressed in colourful tulips, welcoming families for picnics and traditional dance parties. For Afghans, the value of tulips goes beyond simply aesthetic beauty, well past being attractive spots for sightseeing. These flowers are deeply rooted in our culture, they are symbols of the start of a new, peaceful and prosperous year ahead. However, with the rising temperatures, climate science experts are increasingly worried that tulip bulbs are not going to survive this shift to temperatures for which they are not adapted. This concerns us that this will signify further desertification, and will turn our tulip-dressed deserts to plantless ones of only sand and rock all year round.

If world leaders are listening to my voice, and are yet doubtful as to whether climate change is a real catastrophe or invented propaganda, it's understood that they have probably been born into relative luxury in a decorated house, growing up miles away from the daily negative impacts of climate change. It's hard to even try to believe in a crisis that has never really touched you, so therefore, as a climate activist from Afghanistan, I invite you to pay a short visit to where I grew up. Together we will lift the door to a tent of the nomadic people or, as we call them, the Kochi people, of Afghanistan. These are the people most vulnerable to the impacts of climate change in our nation. They live in these tent-like homes and engage in pastoralism as their main source of income. We will listen to the stories of more and more of their animals' untimely deaths, as well as their forced sales due to the severe drought conditions that have meant a lack of vegetation in their feeding spots.

I understand some leaders are angry at climate activists for 'catastrophizing' this process of change, but while climate change may not be the end of the world for the entirety of humanity any time soon, it is ending the worlds of people like the Kochi in the mountains of Afghanistan, as well as residents near the beaches of distant island nations. Where these people meet is where the whole idea of climate justice comes into existence.

This movement is to raise awareness of those in lower-income, less developed places that are dying, losing their homes, their sources of income and their ways of life. It is to protest loudly, in order that we all come to understand that this is in significant part because the very rich, and many big corporations and leaders in more economically developed countries, are denying the crisis of which these oft-forgotten people are the victims.

I was involved in climate activism for about three to four years on a local level, and then, after the Swedish activist Greta Thunberg started the Fridays For Future worldwide movement, I also initiated a climate strike movement in my own country. But I can't expect my country and its government to mitigate climate change on a global scale, because we are not the drivers of this phenomenon. It's extremely important that the communities on the front lines of the devastating impacts of climate change receive the justice that they deserve.

It's very promising, after all, to see that young people around the world, with a wide range of economic backgrounds, are realizing the need for urgent action and they are demanding it – breaking the old norms where people who were the most impacted were not in a position to raise their voices, and where those who were able to be heard went on unaffected and thus remained ignorant.

Planet Earth is home to all of us, regardless of our political affiliation, religion or skin colour. It's time to join hands, listen to each other with open minds and accept each other with our hearts, to save our only home.

NATALIA BLICHOWSKA, 12

England / Poland

Natalia Blichowska (born 2007) would like me to note that although she mainly uses she / her pronouns, she is comfortable with any other pronouns as well. She is an activist in Bournemouth, England, where she has lived her whole life. However, both her parents are Polish, and she frequently goes to Poland, so she was able to provide a dual perspective on climate strikes and the climate crisis.

In February 2019, Natalia joined her first ever climate strike, and subsequently became one of the organizers of the Bournemouth group. By the summer holidays, and her next visit to Poland, Natalia had been an activist for several months, and so she decided to continue striking, outside the town hall in Szczecin. This accidentally ended up setting off seven weeks of holiday strikes, proving that it's not just famous activists like Greta Thunberg who can motivate people to take action.

In November that same year, she was inspired by Vanessa Nakate (whose essay appears later in this book) to start striking for the Congo rainforest, which she does either in school, holding her sign at break and lunch and whenever she gets a spare moment, or online. In fact, she was one of the first people outside Africa to start daily protests for the Congo, and is now involved with organizing the #SaveCongoRainforest campaign.

For a long time, Nat had been aware of the climate crisis. In her essay, she poignantly recounts conversations with her family, back before she became an activist, where they ended up feeling that things just kept getting worse and worse. Luckily, rather than letting the depression drown her, Natalia chose to take action as soon as she realized that politicians weren't going to change, not without pressure.

Climate activism isn't always easy, especially not when you do work behind the scenes. But no matter how hard it gets, Natalia never gives up, on her fellow activists, on her campaigns, or on the hope for a better future.

TWITTER: @e_thunter

My personal experience of the climate crisis is rather limited (that is, I have never faced a 'natural disaster' as such), but I have early memories of watching David Attenborough documentaries and thinking, 'Someone has to do something about climate change!' and discussing with my mum and my brother how the climate crisis could affect our futures... One time sticks out to me in particular:

We were walking home from school one really, really hot summer afternoon. The grass on our school playground was so dry we all agreed it was dead (luckily it wasn't, and it regrew after lots of rain), but the grass that grew in the park outside our school was still quite alive. It didn't keep the heat away from our thoughts though, and soon a discussion about how climate change was already affecting us floated to the surface of the conversation. We started with the heat. Then with how actual snow was becoming rarer and rarer. We talked as we got to the end of the park, and got in the car (yes, how ironic... but it shows how people change! We barely drive these days – and back then, we were convinced that politicians would come up with something) and continued the topic. We said how we were sure that things would get worse and worse. It was upsetting, but we stuck to the topic. I don't remember all that much, but we were in a 'pessimistic spiral' with things seeming to get worse and worse... then my mum turned around and said something like 'I wish I could talk to you about unicorns and fairies and not really depressing things... I'm sorry about the state of the world we've left for you.'

I remember looking at my brother, and thinking how he doesn't really understand this, and other deep topics. So I looked back at her and said, 'I'd much rather know the truth. Lies won't save us.'

She nodded. 'You're right. They won't.'

That conversation has been lodged in my memory to this day.

That was also the summer that Greta Thunberg started her climate strikes.

Coincidence? Obviously. But it shows this was being thought about, that climate change was, and is, on many people's minds. From then-eleven-year-old schoolchildren in the UK, to the icons of our time.

There is so much I could tell world leaders... I could say how furious I am that virtually nothing has been done about the climate crisis (well, apart from big words and promises that seem to be nothing but a way to convince society that those in charge are acting – are they?), especially since they were aware of the impacts of our rising CO_2 emissions... I could say how I don't understand how they kept going, even though they knew what would happen... but actually, I can answer that second one for myself. MONEY. I don't want to jump to assumptions about ALL politicians, and say that they are all greedy boomers, as they are not. There are young politicians, and there are nice and not-greedy ones. There is no rule. But those who 'get to the top', more often than not, are connected to fossil fuel companies, who are responsible for the CO_2 in our atmosphere. I don't want to spread blame and hatred, as that has never been the point of the strikers, but it needs to be said. If you know about the damage you're doing as a person in your life, then you lose the right to excuses. As if one can make excuses in a situation as serious as this, in a matter of survival.

I don't know how long I've been aware of the climate crisis. As mentioned, I used to watch plenty of documentaries, and I can remember discussing with my family whether or not to take the bus more, to eat less meat, all because we were aware of what climate change was doing to our ecosystems and lives. I can remember feeling somewhat hopeless about the issue, and I was very scared about my, and my children's, futures. I remember being convinced that people would act. They didn't. Or haven't yet.

Once I became a climate activist, I started to look at things and think how they could be improved to be more climate friendly, but let's just say the UK wasn't really doing well with initiatives for improvement. Likewise, when I went to Poland for my summer holidays, I saw plenty of potential for things – 'things' as in 'everyday life being greener' – but not much initiative from many leading politicians. What I felt at this point was pure irritation and frustration. Why? Why? Can they? Yes, they can make the world greener. And most politicians don't. Now, this angered me. 'You have answers!' I'd think to myself. But I was also scared that humans, intelligent, smart, world-shakingly brilliantly minded HUMANS, would literally put themselves on a path to self-destruction. And then I was also scared that if we made it out of this mess, an equally inadequate system would be set up... but that's the far-off future. One crisis at a time.

So I associate lots of emotions with the climate crisis: fear, anger, frustration, hopelessness. But I try to stay positive. We can't give up and fall back into that 'spiral of negativity', because things will seem much worse than they are (which is difficult, but somehow achievable by the climate crisis...).Things will seem more daunting, intimidating. We will freeze in fear and inaction if we fall into a negative spiral, because that is just how terrifying this situation is. It's enough to make us freeze and ignore the issue. Understandably. Humans will be in denial of something so terrifying and negative, so to keep us going we must try to see the positive. We might stop seeing the light at the end of the tunnel. I keep saying it: we can't put ourselves down. We must continue. We must be determined and resilient.

I don't really consider myself qualified to give advice about starting climate strikes, seeing as I got involved in climate activism by mistake, but somehow I managed to start seven weeks of strikes in Poland (I was there for five; they lasted until the end of the summer holidays), so here goes:

Remember, people who see you are likely to react with surprise, rather than hatred/disgust. You are more likely to get a positive response... but there will always be people who don't understand, so they may respond by trolling you. There are two things to know in this situation: A) don't explain if they aren't even trying to listen, and B) you will always have support from someone, even if you are striking alone. (There is a Twitter account dedicated to supporting solo strikers – @solo_not_alone.) Someone will always have your back, either digitally or physically, by striking with you. You are never without support.

Know your facts and what you're doing. Have the statistics and have the data. Know the science so you can tell people the truth. But don't depress and overwhelm them, ask them whether or not they want to know the cold, hard facts. If they don't, just tell them we're in a bad place. If they do, tell them as much as they want to know. We can't scare people away.

Strike somewhere noticeable, where people are more likely to see and notice you, but make sure that the people who are most likely to see you are those who are your 'target audience'. For example, if you want politicians to notice you, strike outside the (e.g.) town hall. Likewise, if you want your classmates to join you, or simply raise awareness in school, strike at school during break and lunch, maybe. If you want 'regular' people to see you, perhaps strike in a shopping centre.

Use calendar events to your advantage. For example, the November 2019 global strike was on the twenty-ninth, also known as Black Friday. The square where we had the Bournemouth strike was packed with shoppers! I saw several schoolmates, who ignored me, but they were now aware of the climate strikes. The next global strike was 14 February 2020, so that's Valentine's Day. I felt kinda bad that we may have ruined some romantic moments, but there won't be any romantic moments if humans go extinct.

It's okay to feel self-conscious. You may feel nasty glares focused on you, I know I feel like that, but most of them are probably your imagination, because you expect a negative response. As mentioned, people are more likely to be curious. It's the shape of society… the image of self-consciousness, and that if you're doing something 'weird' then you should feel self-conscious. Well, I hate to disappoint you, but we're not going to solve the climate crisis by playing by the rules and society's expectations. We have to momentarily forget what people expect of us. We can't all be normal. People will, of course, be surprised. No change has ever come without being 'out of the ordinary', remember that. I remember sitting outside the town hall in Szczecin, thinking, 'Okay, I really feel self-conscious, but compared to what Greta felt when starting the movement, I'm probably feeling very little.' (Though having never met Greta, I don't actually know what she was feeling.)

Thank you for reading. My message is keep fighting, stay determined and know what you're doing. We need to take care of ourselves now, like we plan to take care of ourselves in future – and not just ourselves, but also our families. Friends. Colleagues, peers, humanity. If now feels hopeless, focus on what you're going to do once our future is safe. If you feel like you can't focus on you, focus on someone close to you and what you can do together – when you're safe. And act now.

NIA, 19

Argentina

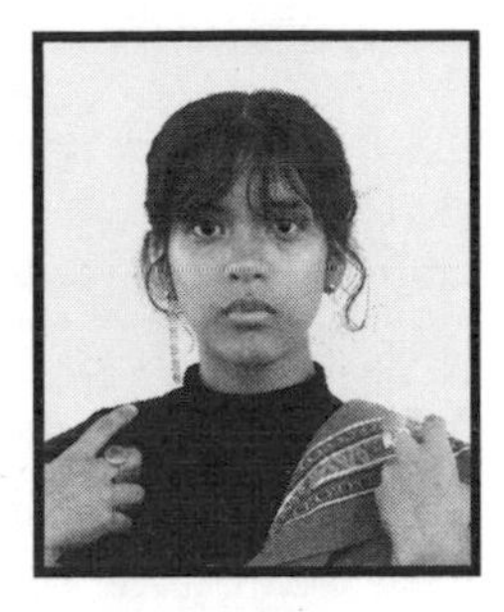

Nia (born 2000) is an independent Argentinian activist primarily focused on Indigenous rights, working with their Indigenous 'brothers and sisters' around Argentina and Peru. Their mother is Indigenous, coming from Chincheros in the Peruvian Andes, and their extended family still lives there, which means that Nia hears 'about everything they are facing as a consequence of the climate crisis. This is something that really moves me.' They are also part of Polluters Out, the international youth coalition co-founded by activists Ayisha Siddiqa (whose essay appears earlier in this book), Helena Gualinga and Isabella Fallahi. They were one of the people who brought that movement to Latin America, where it is called Raíces Libres.

Nia became an activist because they were 'sick of the cruelty that is committed by polluting companies in Indigenous territories and [to] my brothers and sisters, who are actually suffering the worst effects of the climate crisis right now.' They are also keen to point out that, while for many people the climate crisis feels like a recent issue, the problems facing Indigenous communities have been there for hundreds of years.

Nia says, 'I have seen the climate movement mostly focused on white, middle-class activists who just worry about the future, and I feel like we are forgetting that the climate crisis is already here and there are some groups who have been facing the effects for years. I don't like to be in the headlines, because I want to use the space I have to bring visibility to Indigenous groups and [focus on] listening to them (not speaking for them), because this is also their fight and they have the historical right to be head of this movement.'

Nia used to be part of an Argentinian climate strike group, but now they are mostly working independently and with Indigenous peoples.

Nia has been erased many times. They know what discrimination feels like, being part of multiple minority groups. They've been cut out of photos which mostly feature white climate activists and their story has been erased by the media. They want to live in a world where minority voices are also heard. That world, to them, feels like fiction at the moment. But with amazing people like Nia around, I know we can change this.

Hear their message. And join them in their fight for a fair, sustainable world. A world where equality, and a stable climate, are not fiction but reality.

INSTAGRAM: @haluami

Fiction

What is fiction? If you look for the meaning of fiction in a dictionary, you will find that fiction is defined as 'something invented by the imagination or feigned', or 'the type of book or story that is written about imaginary characters and events and not based on real people and facts', so you can describe fiction as something unreal, imaginary. We think about books, about tales, maybe about a movie or a series. However, fiction can go beyond that. What is fiction to you? What is fiction to me? For me, fiction is living in a world of acceptance. For me, fiction is living in a world where people listen to me, a world where my skin colour, my gender identity, my sexuality, my socioeconomic status, do not present a reason to be neglected. For me, fiction is having a voice. For me, fiction is not having to fight every day for my existence.

Now, fiction can also represent something we see as so far away from our realities that it feels imaginary or unbelievable. As an example, for me, fiction feels like every time my mother tells me her story, how she travelled from Perú to Buenos Aires at sixteen years old – alone, with almost no money, leaving her family behind – in order to have a better future, to give me a better future. Fiction feels like every time she tells me how people from our Indigenous community were, and are, being murdered every day defending their lands against companies, governments and extractivism. Fiction feels like every time I realize they are silenced, even though they have been dying for years, because no one lets them speak up.

The world around me feels like fiction.

Sometimes people see me as fiction, too. When I open up and talk about structural racism, white people see me as fiction, because in their reality racism does not exist; it is not visible. They have never suffered because of seeing their lands, their houses, on fire, or officials forcing people to move from their homes because a mining or an oil company from the Global North has decided to start a new project that is going to destroy the places they live in, is going to bring pain to families and workers. They don't know what it's like to see everything that is sacred to you, the crops, the earth, the river, the mountains, being exploited as a resource. They don't know what it is like worrying about the safety of your family every day, because in this society, Indigenous people are not worthy of anything that is not exploitation. They don't see us.

Our oppression is fiction.

When I talk about the climate crisis, and explain how companies are burning our lands, displacing us, ruining our self-sustaining economies, murdering us, persecuting us, how people of colour, mostly people from Abya Yala (the so-called 'Latin America'), Africa and Asia, are going to face, and are facing, the worst consequences first-hand, I realize that almost no one, except us, knows or cares about this, about us: *we feel like fiction.*

All around me, environmentalism, companies, governments say climate change is about the future. Whether they say we have ten years, thirty or fifty, it's always about the future. It's always about when it's going to affect cities, mostly cities from the Global North. They put a blanket over us, trying to hide all the damage, the pain and the oppression. They deny that the climate crisis is about the present, that this is happening right now, they ignore the floods, the fires, the droughts, the destruction of thousands and thousands of lives.

The denial of our realities makes us feel like fiction.

From the city we see the news of the Amazon jungle being burned, and it feels like fiction. The people, the ecosystem, dying is all fiction. We sit on our couches, watching TV, watching this massacre in the comfort of our homes, and it feels like fiction. It is not close to us in the city, so we continue to ignore it. We know it's there, they tell us it's real, but we still can't completely picture it; we continue living in our realities, our little bubbles, without thinking that *this is real.*

This is happening right now.
The world is burning.
We are burning.

But many people still think: 'It does not affect me, so it does not exist.'

I have been told this so many times after speaking up; not only that, but I saw how the media tried to deny my existence, my presence. I was cut from every photo and video from the global strike on 27 September 2019 in Buenos Aires, because my sign was making evident the fact that Indigenous people have been suffering the effects of climate change for years. They decided to silence me because I demanded to be listened to as an Indigenous voice, because, as a non-white person, I asked a white-dominant country, whose history is stained with the blood of Indigenous communities, to make itself responsible for the destruction of our homes and the lives of our brothers and sisters. However, the strike, according to the media, was only attended by white, middle-class activists. They made me feel like fiction.

Their indifference makes me feel like fiction.

But we are not fiction. I am not fiction. We exist. We are here. We are suffering. We are dying. We are being oppressed every day. We are being discriminated against every day. And even though society tries to keep us quiet by making us feel invisible, we are not going to give up.

Our house has been on fire for over five hundred years, since the beginning of the genocide that was and is colonialism in our lands.

We are a reality, a reality that is trying to be hidden and masked as fiction. A reality that is being obscured every day. A reality that is rising. A reality that is going to keep fighting. A reality that is not going to be left behind.

Because we are here, and we exist. *I exist.*

PATRICIA MUMBUA KOMBO, 24

Kenya

After droughts ravaged her country, leading to famine which tore apart her community and took the life of her brother, Patricia Kombo (born 1996), who comes from Makueni, Kenya, realized that she had to stand up for the environment. Until then, she had been a journalist, but with the climate crisis right there in front of her, Patricia knew she couldn't just stand by and watch the crises unfold. It was the beginning of an incredible, life-changing journey.

Patricia founded the PaTree Initiative, which fosters environmental education in schools and is helping reforest Kenya. She also regularly protests on the street, usually with signs related to this aim of greening Kenya. She's already donated and planted trees in several schools, and she believes that this will encourage the children to become climate champions. In June 2020 Patricia was recognized as a land hero by the United Nations Convention to Combat Desertification. She has also received an award from Kotex Kenya for her efforts to promote the livelihoods of Kenyan women.

Her initiative has four main goals. Firstly, to 'foster environmental education in schools through the creation of environmental and "4k" clubs'. Secondly, to 'enable children to take part in greening Makueni by donating trees and training them on reducing pollution, recycling and reusing for [a] sustainable environment'. Thirdly, to 'help the nation attain 10 per cent forest cover by planting both fruit trees and non-fruit trees'. And lastly, to 'educate communities on climate-smart techniques'.

It's a lot to take on, but Patricia knows that change has to happen, and it has to happen soon. Despite the fact that she has been pursuing a degree in communication studies, Patricia has still kept on with the environmental work.

Kenya has been hit hard by the climate crisis. Apart from the appalling droughts and famine, the semi-arid conditions make it hard for people to farm. As a farmer's daughter, Patricia knows exactly what this is like. She also knows that the aridity is caused by illegal logging and the consequent lack of trees. This is a big reason for her actions. She also knows that, in her country, women like her are the most affected by the disasters. I know her fellow tree-planter and countrywoman, Wangari Maathai, would have been proud at how Patricia continues to stand up, fight for what's right, and inspire people everywhere.

INSTAGRAM: @pattykomboh / TWITTER: @patriciakombo

A Tree Planter on a Mission

There was no doubt that in April, rains would come, but the unexpected had happened. I watched with pain as my mother helplessly walked, holding my dying brother by her palms as death snared and patiently waited to swallow us. She gave us this promising smile, but deep inside her, she saw no future in us.

It pains me to this day, seeing my community perish, with nothing I could have done to rescue them from the pangs of hunger. It was in March 2019, and the sun was scorching, and we were experiencing the worst effects of climate change, which had led to prolonged drought, leaving the land unbearable. The land had turned into a desert, with no single trace of a tree. I can say it was the effects of illegal logging that had led into drought, and nature was sad with mankind.

I grew up on the slopes of Kilisa Forest, in Makueni, Kenya. When I was in lower primary, aged eleven, Mbooni, the village where I grew up, was one of the places which were used as recommendations to any visitor, for it was enriched with all sorts of biodiversity. There were lots of streams, and everyone had the weather patterns at their fingertips. We were introduced to farming in the lower classes, with what we called the '4k clubs' – the four k's in Swahili stand for Kenya, *kuungana* (to unite), *kufanya* (to do) and *kusaidia* (to help) – which are now dead. They died because of the theory-based curriculum where teachers want students to cram and pass exams. I miss the sweet organic arrowroot and cassava, which were served with millet porridge for breakfast.

I did well in my examinations, which I attribute to a conducive environment and the healthy organic food I ate, which made me concentrate in my academic studies. My dream was to tell stories, and I was admitted to Moi University to pursue communications and journalism. I had experienced drought when I was eleven years old. I never knew that the real drought would strike, and that this time round, it would be more severe.

In March 2019, the worst happened, the media announced. The government spoke of our dying brothers and sisters in Turkana, who needed our help. Famine had struck again, and it was claiming lives. I wrote an article – 'I Can Feed Myself, Teach Me How to Farm' – which got lots of viewership. I did an appeal video, to aid with the donation of food and material things. I was invited to go to Turkana to deliver the donations, but what I experienced opened my eyes, and an inner conviction whispered to me, 'You were not made to be a journalist but an environmentalist.' As a youth, I knew I was the solution to climate change.

Watching my fellow Kenyans, and especially young ones, die due to the effects of climate change, made me be their voice and train them to act. From that, I founded the PaTree Initiative, whose aim is to educate young pupils and conserve the planet through tree planting. My goal is to help my nation attain 10 per cent forest cover by planting both fruit trees and non-fruit trees, and to foster environmental education in primary schools, through the creation of environmental and '4k' clubs, by enabling children to participate in greening Kenya, and training them on reducing pollution, recycling and reusing for a sustainable environment.

A liveable world and a safe future is our right, and when it is denied, it's our role as young people to rise up, demand and act. It's the time to raise a generation that addresses issues and protects the vulnerable in society.

To my leaders: the world's forests and water bodies are the main life-supporting mechanism for Planet Earth, and they are key to sustaining biodiversity and the world climate system. No matter how much we major in economic growth, it's of no value if we fail to consider climate change.

There is no Planet B, and for the countries to fight and end poverty, we must take good care of the environment and climate, for they hold the future of other sustainable development goals. Eradication of rural poverty is closely linked to the fight against deforestation.

For us to be accountable, leaders should account for the economic value of forests, which are not covered by national accounting, thus contributing to illegal logging and desertification.

To my activists: we are not fighting as individuals. In our activism, we should act. Going on the streets without showing the masses what they need to do will not change things. Organize tree-planting activities and cleaning activities alongside the strikes, for it's a decade of restoration. Enough of climate talks, what we need now is CLIMATE ACTION.

To climate deniers: just like Covid-19, the effects of climate change will not spare the rich. If we reach a point of irreversibility, we will all suffer from the pangs of the climate crisis. Let's join our hands and act, because the climate crisis is claiming lives.

Parting shot: Education is the most powerful tool for change, and a child without environmental education is like a bird without feathers. It's high time we change our education system, and include environmental education for our kids, for, unless they are involved in conservation, they will not value the importance of a better environment.

Better climate, happier existence.

PATSY ISLAM-PARSONS, 19

Australia

Patsy Islam-Parsons (born 2000) is a climate activist from Sydney, Australia. She is the founder of Fridays For Future Sydney, which she started to get more people involved with striking weekly. She is also part of the #SaveCongoRainforest campaign and is a member of the international youth coalition Polluters Out. At the time of writing this, Patsy is studying French at the University of Sydney.

In her country, people have seen the realities of the climate crisis first-hand. In 2019, devastating bush fires swept across Australia, killing billions of animals and destroying countless homes and lives. But Australia continues to be one of the world's biggest exporters of coal, and many Australians, contrary to what you might think, still deny the climate crisis. And Patsy is still often alone in her protests. Worse, she experiences actual verbal abuse from climate deniers when she's out on the streets. Although this is nothing like the challenges faced by activists in some countries, it isn't to be taken lightly, and it's a testament to Patsy's commitment that she keeps going.

Australia rarely has large climate marches. They occur only a few times a year. When they do, of course, they can be really big, and gain a lot of media attention. Sadly, dedicated activists, like Patsy, are ignored despite striking every day, simply because they are often in small groups, or even totally alone.

And of course, being alone is uncomfortable and lonely for climate activists. After all, we're human too. And we have lives outside of being activists – Patsy plays tennis and is a dedicated Tottenham Hotspur fan.

Her message is that we need everyone, no matter who you are or where you come from, because the climate crisis is affecting us all. And if it isn't yet, it will.

Patsy started striking for the Congo rainforest in December 2019. Why? Because, she says, 'The destruction of the Congo rainforest, and the detrimental impacts the climate crisis is having on the people of Africa, are almost completely ignored. Our aim is to change that.' And this thought-provoking essay shows not only that we can change things, but that no one is insignificant in this fight.

INSTAGRAM: @patsyip_ / TWITTER: @Patsyip_

'Recent projections of fire weather suggest that fire seasons will start earlier, end slightly later, and generally be more intense. This effect increases over time, but should be directly observable by 2020.'

Garnaut Climate Change Review (2008)

We all know that there are a seemingly infinite number of so-called 'climate deniers'. They seem to be able to come up with an unbelievable variety of excuses for why climate action isn't necessary, why it's all a hoax, why it's impossible, why it's too late, why it's too early or why it's another country or person's fault. And as we know, they also frequently take to criticizing and trolling us climate activists, mocking our actions, opinions and even appearances and personalities.

But there is also another group of people: those who would describe themselves as 'believing in climate change'. And yet they don't act. They don't seem to be concerned. They aren't on the streets demanding action from their leaders. They just sit back passively and watch. I like to describe this group as those who are stuck in the 'polar bears dying due to premature Arctic ice melt is bad' frame of mind. They have not yet appreciated that we are in a climate emergency that has already completely changed our lives, and which will continue to do so. They don't see that our leaders have our futures in their hands. They don't realize that we are living in some of the most significant years in humanity's history *They aren't scared.*

I am often asked if the 2019–20 Australian bush fire season was the moment that sparked people into action. The moment that changed the minds of deniers, and turned those of the 'polar bear' mindset into activists. I mean, it sounds reasonable, right? Our country was literally on fire. The smoke was so thick that you could taste it. There were days when it was physically hard to breathe outside, and the sky was a mixture of grey, brown and orange. Every morning I would wake up and check the status of the numerous fires roaring across the country. How much land had been swallowed up overnight? How many houses had been burned? Had any more people lost their lives? How many fires were still uncontained? The images coming out of directly affected areas like Nymboida, Mallacoota and Cobargo were beyond horrific. Skies so choked with smoke that it was pitch-black at 9 a.m. Communities forced to take shelter on the water when they received the terrifying 'It is too late to leave' message from the Rural Fire Service. *I was scared.*

On 17 December 2019, the record for Australia's average maximum temperature was broken, with recorded temperatures of 40.7°C. The following day it was broken again at 41.9°C. These record temperatures and the extensive drought fuelled fires virtually impossible to control. They were creating their own weather systems, and the smoke was so great that it was starting to reach New Zealand and beyond. In total, well over 12 million hectares of land were burned, more than a billion animals were killed and at least thirty-four people lost their lives. I asked myself, surely this has made people wake up and realize that we are in a crisis? *Surely now they are scared?* But I'm not so sure that it's had a lasting impact on public opinion. Of course, people who were directly impacted will never, ever forget. If your house burned down, if you were forced to evacuate, if you, your family and friends were in danger or if you lost a loved one, that trauma remains with you forever. The incredible firefighters who risked their lives to protect others will never forget the horrors of what they witnessed. But for the rest of the general public, there has been little evidence that much has changed. We are in a crisis, but hardly anyone is acting like it. *Hardly anyone is scared.*

Just like the people they govern, our leaders still refuse to act. *They certainly aren't afraid.* Some choose intentionally to spread misinformation and divert the attention of the public elsewhere. Others insist on endlessly discussing distant targets that are vague, and fail to specify precisely how they are to be met. Such policies are not plans for climate action. They are plans to pretend to take action. And, in my opinion, that's arguably worse, as it gives a false impression that appropriate and adequate action is taking place. Especially to those still in the 'polar bear' mindset, who are easily convinced that we've got this under control.

So, what do you do when the deaths of billions of animals have virtually no impact on public opinion? What do you do when the government appears unconcerned by the loss of millions of hectares of world heritage forest? Or at least the loss of public property and human lives? We strike. We strike to make them understand our fear. A persistent presence on the streets to show how we feel, every single day. We strike because we know it's possible to change people's minds. I too was once in the 'polar bear' category, so I know that it can be done.

We are asking for our leaders to listen to the science and immediately develop policies that halt all emissions. We are asking that they listen to the experts, and act as any rational person would in an emergency. We are asking that they not gamble with our only planet. Targets won't be achieved by themselves. We cannot address this crisis without making drastic changes to almost every aspect of society. We've been given enough empty words and forgotten promises. We need action.

Now is the time that we need everyone to join us. Every person who is reading this, we need you. We need your family. We need your friends. We need your neighbour. We need the colleague who you have lunch with. We need the guy who sits next to you on the bus. We need the waiter at your local cafe. We need your child's favourite teacher. Each and every one of you. We need you.

RAYMOND SIMPIE SMITH, 17

South Africa

Raymond Smith (born 2003) is a climate advocate from Glencoe, KwaZulu-Natal, South Africa. He isn't really involved with big groups like Fridays For Future, but that doesn't mean his work isn't important. He aims to educate people about plastic pollution and what it can do to our oceans and to the environment. He also raises awareness on social media, telling people about these issues so that they can act. And he takes part in voluntary beach cleans in his area. He is also starting a climate group with some of his friends, called Us to the World.

Raymond says that he wants to do more, but it is hard to do so in his country because many organizations don't let people volunteer or donate unless they are at least twenty-five, meaning that he has mostly to 'go it alone'. For instance, he wants to help Greenpeace Africa clean up Durban oceans, but has never really received a reply confirming whether or not he can join them.

Raymond started getting involved with climate issues in about 2017. Coming from 'one of the cleanest towns' in South Africa, he only started to realize the impact of plastic pollution when he visited other places. That was when he started raising awareness. He was inspired by two people. The first was his teacher when he was thirteen, who was an environmentalist and therefore taught them about climate issues even though it was not part of the curriculum. And the second was Beyoncé, who inspired him by going vegan.

Raymond's goal is 'to educate as... [many] people as possible about the consequences of pollution and how a small piece of plastic can do so much harm'. He finds all climate issues equally important, and fights for all of them equally.

Having experienced droughts as a consequence of the climate crisis, Raymond knows that this is absolutely real, and that, in places like South Africa, it will hit hard. So he keeps going.

Just as one piece of plastic can change things for the worse, one person, like Raymond, can be a change for the better.

INSTAGRAM: @smith_raymonddd / TWITTER: @raymond_smithr

As we all know, climate change has plagued our Earth. It was pretty much inevitable, but certainly could have been controlled and maintained. Rising temperatures melting the glaciers; thousands of animals losing their homes and dying out because of their inability to adapt to these new environmental conditions. High rising sea levels and ocean temperatures… all these atrocities, and what are you doing as an individual to help?

The last time I saw snow was in 2012, exactly eight years ago. The detrimental effects of global warming have left us with brutal ramifications, as we've been left with aridity in some parts of South Africa over the years.

Climate change has aggravated soil erosion in most areas, making it impossible for plant biodiversity to survive, which in turn impacts local communities and the economy, as this affects the farming industry and food security.

South Africa has experienced several droughts, the recent one in northern KwaZulu-Natal coming to a halt temporarily in the last quarter of 2019 as we received generous amounts of rainfall. This has affected our local farmers so much because of biodiversity loss. The results of this dilemma made it difficult for farmers to plant, and to maintain livestock. With that being said, it had dire consequences for the economy, impacting it negatively.

Over the years, in South Africa, the intensity of climate change has affected the quality of the water supply and there is scarcity due to the subtle changes of precipitation patterns in the country. The country at large has suffered droughts since 2015, resulting in crop losses and water restrictions for everyone. How would you feel if you had a strong urge to freshen up, and stepped in the shower only to come to the realization that your water had been cut off due to water restrictions in the area? Awful, it is.

These changes in our climate have also resulted in floods because farmers had to destroy wetlands to make space for crops, and essentially wetlands act as sponges during floods.

These are among some of the many personal experiences I have had in South Africa in the midst of climate change.

To world leaders and people in higher positions, those who command power, I pour out my heart to you to please assist climate activists, as we attempt advocacy for our mother planet. Please fund environmentalists to successfully protect our world. We need your cooperation to make our Earth a safe haven for future generations, and our present society. We're fighting a losing battle, being the minority of people who are trying to save the world. Thus, I appeal to you to give climate change activists the platform to preserve and conserve our Earth. I appeal to you to amend the legislation regarding the environment, ensuring full protection of our environment, and to enforce these laws rightfully. I ask of world leaders to stop with the wars, as nuclear power destroys the ozone layer. We can surely all live in peace. I ask of world leaders to fine anyone that litters, to invest more in recycling and reducing. I ask of world leaders to please ban plastic straws and reduce the use of plastic in general.

It brings me great sadness and despondency to see our Earth slowly die like this because of arrogance and ignorance. It pains me to see how much the world is uninformed of the great consequences of climate change, and the ways we can all save the planet. As an environmentalist, my biggest issue has always been the ignorance in society, how everyone has been indoctrinated to believe that just throwing a plastic wrapper away has little to no effect on the environment. It hurts me to see how people litter and see no problem with that, and me having not much authority to make them pick up their litter.

I have hope and I have faith that the world will heal and it will be glorious.

The many other environmentalists and climate change activists out there, you are all doing a sterling job, and it's commendable. No matter how small you think it is, you're impacting everyone around you in a positive manner. I want every environmentalist to believe that they are changing the world, to continue doing the amazing work and to continue fighting for our planet. The world is a large place, but baby steps at a time can really go a long way, so we are capable, and are indeed changing the world.

I dedicate to you all 'I Was Here' by Beyoncé Knowles-Carter, to empower your inner fighting spirit and to propel your diligence.

RITA NAUMENKO, 16

Russia

Rita Naumenko (born 2003) is a climate activist from Russia. She was born in Murmansk, but then her family moved to Moscow. Rita is part of FFF Russia, and strikes regularly.

She first became involved with climate issues in spring 2019. She volunteered at a public lecture, 'Education in the Era of Climate Change', and after that she became interested in environmental issues. Before that, Rita barely knew what climate change meant, because of the lack of information about it in her country. But after this, she started to read articles and scientific papers about the climate crisis, and to change things in her own home.

It was through social media that Rita discovered, to her surprise, that Russia had a Fridays For Future group, and that they were holding a mass strike in Moscow, in July 2019. She says, 'Without thinking twice, I decided to go there, because I felt the importance of the climate crisis problem.'

After this, she didn't look back. Even though it is hard to be a striker in Russia, and even more so to be under eighteen and a striker, Rita still protests. Even though she has often been brushed aside because of her age, or told she needs her parents to speak for her, she still keeps going, because she knows that the climate crisis cannot wait. Even though she usually has to strike alone, she doesn't stop, because she knows so many stand with her, some in Russia, some elsewhere.

Although Rita's age has often led to difficulties for her, she knows that being sixteen doesn't make her voice any less important. Because young people are perfectly capable of thinking, speaking and acting for themselves. Just like Rita has. By standing up, by refusing to be crushed, by speaking out with courage, determination and perseverance, Rita Naumenko is making sure that she is heard. And, if we manage to keep global heating below 1.5 degrees, I know Rita's name will not be forgotten in a hurry.

INSTAGRAM: @paaiegen / TWITTER: @paaiegen

'As I know, there is a minor here. Someone is seventeen,' said Ruslan Edelgeriev, Vladimir Putin's special representative for climate issues.

'I am sixteen,' I answered, without a second thought that it might be a problem.

'Then where are your representatives? Your parents? You must come to such negotiations with the person in charge. Okay, next time, keep this in mind.'

My name is Rita Naumenko, and, as you will have already understood, I was then sixteen. I was born and raised in Russia, a country with many restrictions and limits. The situation described above happened in February 2020, at a private meeting of four Fridays For Future activists with our president's climate adviser. In fact, this meeting wasn't official. Therefore, in theory, I wasn't obliged to come there with adults. Edelgeriev's words made me think about the barriers of youth activism, but, for some reason, I even perked up.

How did I get there and how did it all start? I clearly remember the moment when I heard the phrase 'climate change' for the first time. It was in the spring of 2019. It didn't happen at school, and it wasn't my parents who told me. Unfortunately, I can't even tell you that I learned this several years ago, before this topic was discussed so actively. I started thinking about it after a public lecture by a Finnish social foresight specialist, where I was volunteering. I went there not because I had an interest in climate. Honestly, I didn't have any idea what climate change meant. Sometimes I'm horrified by how many people still don't know anything about this problem. When I learned about it, the concrete wall in front of me seemed to collapse. I had a feeling that something very important, which I should have known, was hidden from me for my whole life. It may sound weird or implausible, but such human ignorance is easily explained. The media coverage of the climate crisis is really poor here. Only a few journalists write about it. It's challenging to find a book related to this problem or to get relevant data. In Russian, I mean. In English you can find literally everything, but not all people have the privilege of knowing another language. So, this lecture was a turning point. I remember that, after it, I went out with shining eyes, and firmly decided for myself that I would definitely plunge into this topic.

The first month of summer I spent reading articles and research papers, taking an online course and watching movies and lectures on climate change. I've never felt so enthralled before. At that time I didn't think that this topic could be not only studied but also promoted in our society, until I came across an announcement of the Fridays For Future mass action on a social network. I hadn't heard a lot about it, but I was pleasantly surprised that such a movement existed in my country.

Now, let me tell you a bit about how the system of protests works here. To hold a mass action you need to fill in a notification form, which includes organizational information, and send it to the administration of the city. They can approve it, or decline using stupid excuses like 'the picket will lead to disruption of the functioning of transport and social infrastructure and violate the rights and interests of citizens' or 'on this square at the time specified by you in the notification form, this previously planned cultural event will be held'. What do you think the administration of Moscow usually does? They reject our notifications about actions for fifty people, although another protest, in which 3,000 people participated, was held in the same venue. It turns out that the organization of mass events in the good venues (not in the depths of any park where no one will see you) becomes almost impossible in Moscow.

However, sometimes mass pickets still pass, but not in the most public places of the cities. The first action I participated in was held in a square on which there was not a soul, only climate strikers. I was upset that this protest got so little attention, but, at the same time, happy about all the people I met there. Next week there was also a mass strike in the same place. But these two times were apparently exceptions, because after them we didn't succeed in mass actions coordination with the administration.

Someone might ask: 'How can you still engage in activism?' And I'll answer that single pickets are a part of our solution. But another problem arises here: the status of the person on a single picket isn't defined. This person might be considered either an organizer or a participant of the picket. There's no age restriction for the participants, but an organizer must be at least eighteen years old. This inaccuracy of the law is usually used against minors, because in most cases they're recognized as organizers and get detained by police. You don't need anything besides courage, willingness to answer questions, and a passport that will prove that you're at least eighteen years old, to stand alone with a poster on the square. And I have only two things from this list. Despite this, I still go on the single pickets and do it as cautiously as I can. Sometimes the police check activists but, so far, I've been lucky all the time.

It's not easy to be a climate activist in Russia, because of all the obstacles you face. Sometimes I feel hopeless and frustrated, especially when I see how the strikes in other countries go. But the current situation also motivates me to speak more about the climate crisis. Not much time has passed since I became part of the movement, but I can say with confidence that I have changed a lot. And I'm pleased to notice that people around me are also changing. Compared to the end of spring 2019, the community of activists has become bigger and the climate crisis topic has begun to be covered much more. I endlessly appreciate the people with whom we manage social media accounts, organize lectures and participate in events. I also joined a project that focuses on enlightening people about climate change, its consequences and the actions that we can do right now to mitigate it. Now I not only participate in the strikes, but also take part in creating a podcast, which has really become an inseparable part of me. *CO_2-free* is a podcast defining climate change topic by topic, discussing ways to reduce our environmental impact and learning how to adapt to a changing world. We create it in Russian and publish episodes on eleven platforms, including Apple Podcasts, SoundCloud and Spotify.

It was 6 February 2020. I found myself in the climate adviser's cabinet, thinking about my way in activism, after his words about my age: 'Where are your representatives?' This question began to repeat itself in my mind over and over again. Do I really need representatives to fight for my future?

ROWAN WALTERS-BRUNT, 18

England

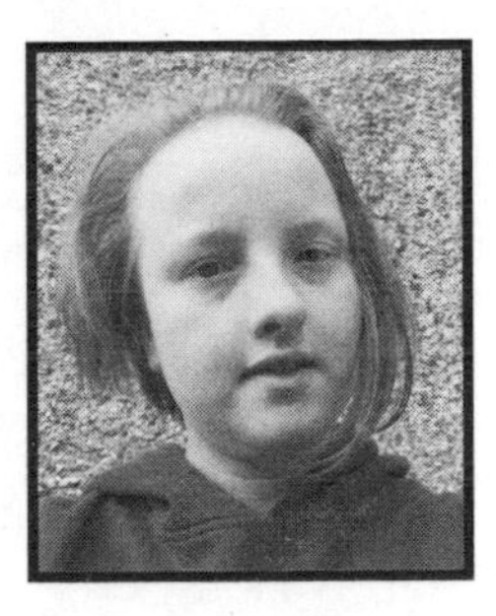

Rowan (born 2001) is a non-binary climate activist from England, and a YouTuber. Xe is primarily involved with XR, but also joins the climate strikes. Xe is particularly active with issues surrounding education, as related to the climate crisis.

Rowan got involved with the climate movement back in April, after seeing the IPCC report and being inspired by XR's 'low-key' actions that had occurred a couple of months before.

In October 2019, xe was arrested for taking part in the International Rebellion with XR in London. The police searched Rowan because xe appeared nervous, and this was apparently 'suspicious' even though xe has both autism and anxiety. Xyr description of this experience isn't easy to read. But it tells a truth that we all need to hear. And sometimes truth is uncomfortable. In fact, it is for this reason that leaders have lied for so long. Truth can hurt, can push you out of your comfort zone. But truth also pushes us to change. Truth is powerful. Truthful education is powerful. And Rowan has the power of truth behind xem.

A while after Rowan expressed interest in this project, I realized that I had actually met xem at the London Rebellion, and had heard a very inspirational and true speech by xem when we marched on the Department of Education.

*The theme of this march was 'Teach the Truth', because there can be no action without awareness, and awareness comes through education. It was a very empowering experience. Though it poured, that didn't get us down, we danced in the rain, and I remember feeling so proud of the people I was with, and of how indomitable they were. And I was so pleased to discover that Rowan was part of that, and that xe was one of my fellow speakers of truth to power. As I shouted into the mic then, '*WE ARE THE TRUTH!*' and no one is more so than Rowan, who isn't afraid to say what xe knows needs to be said.*

TWITTER: @VoidpunkXenon

Lessons Without Lessons

The decision to skip school comes with a particular flavour of scorn from commentators in the media. Back in February, there was an especially hilarious incident where people literally started complaining about a protest being on Parliament's lawn, and apparently this is counter to our values as people who care about the environment. (Never mind that lawns are an ecological dead zone – that's not really the point.) At every major strike or whenever Greta Thunberg speaks, the same clichés are repeated:

'It's just an excuse to skive.'

'You're missing your lessons.'

'Why don't you study to be a climate scientist?'

These statements aren't real arguments, obviously. They just want us to shut up and not disturb their comfortable privilege. But the claim that we're not learning anything is interesting, and demonstrates the fairly narrow view of education we hold collectively.

Most obviously, depending on what role you take at a protest, you can learn organizational skills, public speaking, de-escalation, media, your legal rights. Protesting teaches you ways to engage with politics and power. That's just the obvious stuff, and stuff that you might be able to transfer to formal employment.

But learning is more than that. When you are in school, you learn more than just the written curriculum. You also learn a set of values. These values reflect what society considers the ideal citizen. And those values are competition, individualism and obedience. Once you reach secondary school, you become aware that exam results are determined by the percentage of people who get your marks. The top 10 per cent will get an A*, the next 10 per cent will get an A, and so on. This leads you to hope that your peers will do worse. You know you shouldn't, you want everyone to succeed, but you know other people's success will decrease your chances in life.

This is a side effect of capitalism, given that that is basically how you compete for employment.

So what values do you learn at a protest? The first is solidarity. At a protest, you're all united by a common interest. In this case, making sure the planet doesn't burn. If one person succeeds, you all succeed. So there's an incentive to make sure everyone is okay. If the police start harassing someone, you back them up. If someone seems unhappy, you check in with them. At an open mic, everyone gets the chance to speak. Typing these things out makes them seem simple, obvious and basic, but in other parts of society, we're incentivized to only care about ourselves.

The other main thing you learn is how the authorities handle, sometimes even minor, criticism. The usual response to climate strikes is condescension, but I've also seen outright hostility and anger towards us.

In October 2019 I was part of the International Rebellion with XR. On my way to a site, on the first day, the police stopped and searched me, under the reasoning that I was nervous, stammering and didn't make eye contact. This was apparently suspicious even though I'm autistic, suffer from anxiety and was suddenly being questioned by the police while alone in an unfamiliar city. They found three tiny bottles of fake blood (i.e. the little make-up tubes you get for Halloween) in my bag, and arrested me. The police were perfectly polite and respectful, but in some ways that made it worse. No matter how nice they were to me, they still handcuffed me, and locked me up in a police cell where there was a camera pointing at the toilet, while I did my best to avoid crying in a place where a camera could see my face.

There were also the little moments of disempowerment. While I was being booked into the police station, my mum rang my phone. All I wanted at that moment was to tell my mum I was okay, and to tell her I loved her, but I had to watch the police officer turn off the phone. I knew I wouldn't be allowed to hear her. I was technically allowed to contact her, but didn't want the police listening in, so I didn't. When I was let out, one cop told me that if I went back to a protest site, I'd be arrested immediately, which I'm pretty sure was a blatant lie to try and intimidate me. I did go to protest next time, because I was there to yell at the government, and the police had no right to stop me.

In response to me having a lawyer's number written on my arm, the cops commented, 'They all have them written down, they tell you what to do if arrested, but they don't tell you not to get arrested,' and laughed. This I found very dismissive, especially given the flimsy pretences of my arrest. Even if you believe that the police are just following the law, the law does not call for them to be dismissive and rude. And I couldn't argue back against them because I didn't know how that would affect my case.

I was put in a position where I could not argue my corner. I'd been deprived of my right to freedom of protest and of speech, and privacy. I still constantly thought about it months later. By the time I was released, I was too exhausted from stress and anxiety to challenge the police on whether or not my rights had been respected.

Later, at the climate strike on 29 November 2019, it seemed to me that the police regularly lied and harassed protesters, in the hope of intimidating the less experienced. Fortunately, this failed, given that a lot of us did, in fact, have experience.

These experiences taught me some very hard lessons. One, the police can be perfectly lovely while depriving you of your human rights. Individual niceness does not excuse structural injustices. Two, the police value an abstract order over the well-being of the community the order is supposed to serve. Three, the government really doesn't care about people who don't have millions of pounds, unless you make them. And I couldn't have really learned these lessons in a non-abstract way if I hadn't gone and protested.

I'm not devaluing traditional education. I understand that it's underfunded, and that the teachers are overworked. What I'm saying is that there are lots of different ways to learn, and that learning experiences are not worthless just because they probably won't help make money for a hypothetical future employer. And we're going to use these lessons to build a better world.

SAOIRSE EXTON, 14

Republic of Ireland

Saoirse Exton (born 2005) is a climate activist from Limerick, a city in the west of Ireland. She is a local, national and international organizer with Fridays For Future, and she strikes every week. Saoirse has previously worked with Polluters Out Ireland, the Irish branch of the international youth-led coalition, and a climate science NGO. She is also a member of the Irish Youth Council and is the Equality Officer with the Irish Second-Level Students' Union, and within these she tries to raise awareness and speak up about the climate crisis among her fellow youth.

Saoirse has helped organize three major strikes in her city, and spoken at marches in Dublin, the capital of her country. She has received a few accolades, including being named Limerick Person of the Month for June 2019.

As an international organizer, Saoirse has taken part in Europe-wide meetings and protests, meeting and networking with fellow activists from many countries. Although Ireland hasn't been affected very much by the climate crisis compared to many places, it has seen record-breaking floods three times over the last eleven years, and the government is still not listening to its people. It's the same problem as everywhere. The ones with the power are ignoring those without it. And if we didn't have people like Saoirse, we'd have no one to challenge that.

Luckily, we do, and she's not planning on giving up any time soon. Not while she still has to lie awake at night fearing for what the future will be, fearing for her life and the lives of future generations. Ignoring the teasing she gets from her classmates, Saoirse carries on.

Growing up in a relatively privileged country, Saoirse says her biggest goal is climate justice for all, no matter what. Because we all deserve a better world. One where we can live safely, breathing clean air, surrounded by our essential animals, with adequate water and successful crops. One where the world's youth can look forward to a bright future, rather than living in fear of the years to come. This seems so far away, but by inspiring others to 'take action and to stand up and fight back against all the odds', as she says she hopes to do, Saoirse Exton is bringing it closer. One strike, one fight, one word at a time.

INSTAGRAM: @saoirse_exton / TWITTER: @saoirse_exton

The climate crisis is the biggest issue that faces us today and it is not being taken seriously. I have been climate-striking every Friday for over a year, and during that time I have met amazing and inspirational people from across Limerick, Ireland and the world. I am inspired especially by the adults that join us, every Friday and on the big strike days.

I have been nominated for Limerick Person of the Year 2019, and have been awarded Limerick Person of the Month for June 2019, which means that some people are beginning to recognize the fight for climate, which I am proud to represent in my city. I have co-organized three major strikes in Limerick, as well as speaking in Dublin and at the 5 June 2019 anti-Trump protest. I was a Youth Assembly delegate, where I got to go inside the Irish Parliamentary Houses (the Dáil) and I was an Irish representative for the International Fridays For Future meeting in Lausanne, Europe, where I met and talked with amazing and inspirational activists from across Central and Eastern Europe. I am regularly featured in local media, and I have been involved with national and international media too. I am a member of Ireland's Youth Council (Comhairle na nÓg) and the Irish Second-Level Students Union, where I strive to put climate on top of the agenda when it comes to schools and school-related issues like mental health and the curriculum. I have been involved in Fridays For Future Ireland since March 2019. None of these opportunities would have been possible if it weren't for my activism.

Although climate change doesn't affect Limerick or most people in Ireland an awful lot yet, we do suffer from climate injustice and serious ecocide. Our Environmental Protection Agency refuses to listen to its people, and our Climate and Environmental Office refuses to accept that climate breakdown is actually a crisis that needs to be dealt with as soon as possible. The government has declared a climate emergency and they have a climate action plan, but it is nowhere near what needs to be done to get in line with the Paris Climate Agreement.

Although these developments could be seen as progress, they feel like things that have been put in place just to placate the youth climate activists around the country. Storms are getting worse, flooding is getting worse – in 2009, we had the worst floods in decades, and the same levels of flooding have been experienced twice since then, including this year – but nobody seems to care.

I am teased in school. I am called names and I am asked every single day, 'When are we going to die?' or I am told that I'm just too serious, I'm too panicky. 'Why do you even care?' my schoolmates say.

I often lie awake at night, because my anxiety is like a tidal wave. It hits me smack bang in the face with all sorts of things. Like other people my age, it might be a maths exam tomorrow. It might be homework due, or it could be a decision to be made on whether I need extra tuition, everything caving in my back like a ton of bricks. But, unlike other teenagers, the thing that makes me most anxious is that within the next ten years, we will be beyond the tipping point of climate change. That is what makes me toss and turn and have nightmares. That is what makes me terrified and ready to change, because otherwise, I don't know what I'll do with myself.

Before I started taking action, I used to suppress my fears about the climate crisis, because it is so terrifying. I pushed those feelings so far down that it ended up seeming that I didn't care at all. It was another thing in my complicated tween life. I really didn't realize just how bad it all really was. But then one day in late 2018, I was doing my homework, and my mother showed me some pictures of the huge protests happening around the world, on social media. It was different, because I had never seen kids actually exercising their voice before. So when I saw these protests, I knew I had to act. I decided that I wanted to get involved. I *had* to get involved. So I did. I sat down to protest, in early 2019, and became a part of the movement that has changed my life in so many ways.

Despite all the problems that we have to face as activists, I believe that they actually help to propel us. They help us to realize that despite what we face, we have to continue to stand up and fight back. I am really lucky because I am so privileged. I live in a country where my voice can be heard, where my gender and race don't restrict me, and where I am allowed to protest. Because I know that privilege means that I have a voice where others don't, I strive to put those who aren't as lucky as me at the forefront of what I believe in. We have to put people before profit, or else human civilization will end as we know it. I have hope, and that hope empowers me.

My message to the politicians: companies pay you money just to shut you up. They pay you money so they can go on living too comfortably, while we are all dying. They have led you to believe you do not have enough money to solve climate change, yet fossil fuel subsidies amounted to fifty-three times the Paris allocated monies for poorer countries. Or, if we slash international military budgets by 25 per cent, we could potentially raise 325 billion US dollars. They have led you to believe that overpopulation is the problem, that there isn't enough to go around. Yet, in 2015, sixty-two individuals had the same wealth as 3.6 billion people. Stop with this frankly stupid mindset, and look around you. Imagine this world flooded, trees gone, fires burning everywhere. I want you to imagine your terror at the loss of entire nations to the sea, or entire peoples to fires, or the loss of all growing plants. Now think of your children. Think of me. Stop acting like incompetent children while the children are acting as you should be. You showed us during the coronavirus crisis that we could shut down to save lives, and yet we continue to power on, business as usual, when it comes to the climate crisis. It is possible to lock down. It is possible to stop the use of cars and private vehicles in excess. It is possible to say no to corporations. To tell them that tax havens won't work any more. It is possible to obtain climate justice. If you just try.

I hope that I have given you a small window into my life as an activist from Ireland. I want my experiences to inspire others to stand up and join in with the worldwide movement, a movement that has grown from the actions of a few small individuals into one of the most powerful expressions of youth engagement the globe has ever known.

SOLIS STELLA, 20

Germany

Solis Stella (born 1999) is an environmental activist from Germany. She is involved in the support of renewable energy and is working on a project about environmental psychology. She is also a Fridays For Future activist, with FFF in general and FFF Germany, mostly involved with the organization of teaching about the climate crisis, via lectures or discussions. Within FFF Germany, she works with Students for Future, a group of students fighting for our planet, and Christians for Future, one of several faith groups that combine their religion with their love of the planet, and are, to quote a slogan commonly used, 'united for our sacred Earth'.

Solis, not content with acting in one way, or being involved in just one area, is also part of Psychologists for Future, a group of people involved with psychology using their knowledge for environmental reasons.

Solis has been lucky enough to live in a privileged country, where the impacts of the climate crisis have not yet been fully felt. Or at least, where few people realize how many deaths are actually caused by it. However, this didn't stop her from waking up to what a disaster our current lifestyle has created, nor from trying to find a solution. And it hasn't dissuaded thousands of young German people, who march on the streets every week, in some parts of her country, on an incredible scale, leaving activists everywhere wondering how, exactly, Fridays For Future Germany manages it!

No matter how hard, uncomfortable or difficult she might find it to speak up, Solis is doing it anyway. Because she knows that climate breakdown will not wait. Climate breakdown doesn't care about how we feel. And climate justice will only be achieved if we start thinking beyond here and now, thinking outside just ourselves.

Solis' work might not be centred around intense direct action, or huge protests, but environmental education is equally important. If everyone was taught about the crisis, they would act, so her work is vital if we want to unite for climate justice.

TWITTER: @SolisStella1

The main factors that killed people in Germany over the last decades were old age, accidents, cancer, heart problems and suicide. Despite growing up in a rich country that is usually considered stable and secure, I have been aware since I was a child that some deaths cannot be prevented. I was also aware that extreme weather events start to appear more often over the years because of man-made climate change.

I tried to solve the problem of man-made climate change, and searched for a solution to get CO_2 out of the atmosphere without a not-yet-invented technology. I came up with planting trees. Later, as a teenager, I supported renewable energy and education about the environment, projects about this and related topics. Meanwhile, there were movements believing that there is no man-made climate change and others who believed it but seemed to always put it behind some other issues.

I strongly believe that one core problem about the climate crisis, really, is that people do not know things. I believe there is a large lack of knowledge in two areas. One is between the knowledge of the existing problem and the knowledge of how to act against it. People are aware of the existing problem of climate change, but have no, or not enough, knowledge of how to act against it. I once read the theory that this is the point where Greta Thunberg becomes relevant, because she is an influential figure who deals with that problem by simply giving people a time and a place to be.

The other point is especially about leaders and climate destroyers. While there are multiple factors involved in their decisions, I believe that some are particularly important for approaching the role of climate destroyers: the diffusion of responsibility and the 'distance' from the victim.

According to one German newspaper, 6,000 people died as a consequence of climate change in the summer, in Germany, in

2015. And they used a report from the government as the source for this number. Diffusion of responsibility in this context means that people do not act, or do not act as much as they could, because there are other people in the same situation. It is the classic 'but other countries have emissions too' argument, as well as the 'activists should lower their own emissions and/or get other people to do so' argument. But neither argument means that you should not act.

The second one, the aspect of the distance from the victim, becomes relevant when you look at the consequences of harming the climate. If it becomes dangerous to a rare species on another continent, or to people far away, it is easy for leading people in politics and business to distance themselves. There are actually lists of factors that make it harder for people to distance themselves. If the ones endangered are people instead of plants or animals, if they live nearby instead of living far away, if they are perceived as being similar to themselves.

And yes, I live with the belief that not all, but some, of the climate destroyers would act against the crisis if they were confronted with the consequences more directly.

And I? When I read about the deaths I mentioned, I think about people who I knew who have died or were in danger. I manage not to get stuck on the question of who might still be alive if there was no man-made climate change, and instead focus on the perspective that my activism might help someone – even if signing petitions for more solar panels or standing around in the rain with my sign does not look that way. A strong belief that I should be involved somehow, at least try to help, is an important feeling I have towards the climate crisis, connected to sorrow and anger.

To people who are activists or might become activists, there are many things I would like to say. But particularly important is this: have in mind why you do this. And yes, there are different reasons. Some people are extremely motivated to help protect a rare species which many people have never heard of. Others do it for their hypothetical great-grandchildren. And some for somebody or something they love, a friend or a family member, an animal species or a landscape. And that is OKAY.

There is also something else which I would like to share, for those who believe that they might make things worse. As a teenage author, I did not like my own writings at all, and I continued writing with what I call the belief that 'what I have to say is more important than the mistakes which I will make saying it'.

TESS NORTHCOTT, 16

Australia

Tess Northcott (born 2003) is a youth striker and conservation advocate from Sydney. She has worked with School Strike 4 Climate Australia, particularly the North Shore local team. Tess helped organize the first two global strikes in Australia, and has also spoken at and helped to organize local strikes.

Ever since she was a young child, Tess has been fascinated by the natural world, and this has now developed into a passion for wildlife photography and climate / environmental activism. Even when she isn't protesting, Tess raises awareness online. Her Twitter account is filled with pictures of beautiful and brilliant animals and plants. But these same species are vanishing due to the climate crisis. This is what Tess is trying to get people to understand, through her amazing images and powerful words: the sixth mass extinction is destroying vital links in ecosystems and food chains, and stealing the beauty and diversity of this wonderful planet.

Having got involved with School Strike 4 Climate Australia in October 2018, Tess says she has 'been lucky enough to witness the Australian youth climate movement grow from a small grassroots organization to one that spans across the country, with hundreds of thousands of students striking across the country for the same goal'.

Tess has experienced the effects of the climate crisis first-hand. The devastating Australian fires blotted out the sky in Sydney for weeks, leaving the air quality on a par with New Delhi, and destroyed thousands of hectares of land across the country, including part of the town where her father grew up. And, when Tess visited Indonesia in January 2020, she witnessed the terrifying floods in Jakarta, the world's fastest-sinking city.

If we want to avoid further appalling disasters like these, world leaders MUST listen to the millions of school-striking children. And that means Tess Northcott, because she is one of them. And every single person is vital to the fight.

TWITTER: @Tess_Northcott

In November 2018, I joined over 10,000 students and young people to strike in the streets of Sydney. We filled the city centre, Martin Place, and overflowed into the surrounding areas. For hours, we chanted and called on our politicians to use their power to aid in the reduction of emissions and invest in our country's future.

For as long as I can remember, I have heard the warnings from our world's top scientists, around the globe. In fact, these very warnings are the reason I became engaged in the climate movement. I could no longer bear to hear these dire messages and watch them be ignored by the very people with the power to solve the crisis.

I first became involved in the climate action movement in late 2018, when I read about Greta Thunberg's school strike and found out about the subsequent Sydney strike. I joined School Strike 4 Climate when the organization was very much in its infancy, and I have been fortunate to see it grow from a small group of passionate individuals to a countrywide network of thousands of students all fighting for a common goal. I have been lucky enough to assist in organizing small local actions, and have spoken on behalf of the Sydney branch in the lead-up to the global strike on 20 September 2019, which drew a crowd of almost 80,000 people in Sydney alone.

Since late 2018, more than a million students have participated in school strikes across the world. The combined power of youth from every corner of the globe has begun to make waves in the media. The youth are fed up with watching endless political debates in their home countries, and inaction at a global level in United Nations conferences. We are sick of governments and businesses putting profit above people. Although 195 nations signed the Paris Climate Agreement, and committed to rapidly reduce greenhouse gas emissions, aiming to keep the global average temperature well below 2°C relative to pre-industrial levels, we have seen little action.

Many countries are nowhere near achieving their targets. Much of the public of voting age are consumed with small benefits such as tax cuts, not realizing that inaction on the climate crisis will cause future global financial ruin.

The climate crisis has begun to severely impact Australians in ways that much of the general population has never before experienced. The smoke haze that blankets my home city of Sydney as I write is a constant reminder of the drought and bush fires that are crippling the country. My city has implemented water restrictions, and in January 2020 metropolitan dam levels had declined by 30 per cent since 2013. On top of this, the Great Barrier Reef, the world's most extensive coral reef system and one of Australia's most significant natural wonders, has been steadily declining. Much of the damage has been caused by severe weather events and ocean acidification fuelled by climate change, having a detrimental effect on the longevity and health of the reef. I was lucky enough to visit the Great Barrier Reef when I was eight years old, and can clearly remember seeing the damaged coral, devoid of any life.

In March 2020, the reef experienced its third mass coral bleaching event in five years, following the catastrophic bleaching that occurred in 2016, which killed approximately half of all the shallow-water corals in the reef. Continued inaction on a governmental level will only see these events worsen and become more frequent. Australia faces losing this incredible world heritage site and the economic benefits that accompany it, adding to the future financial repercussions that will be directly caused by the climate crisis.

The summer of 2019/2020 brought extreme bush fires to much of Australia. At least thirty-four human lives were lost, along with 3,500 homes, 18 million hectares of land, and an estimated one billion animals killed. Our cities were choking on smoke so thick that I could barely see the end of my street when I stepped outdoors. At that time, 350 miles away, a rural town called Glen Innes had been strongly impacted by a powerful fire that had already taken two lives. My father was born and raised in that town, and much of the surrounding areas that he once knew are now unrecognizable.

My passion for wildlife and nature is at the forefront of every action I take in regard to the climate crisis. For as long as I can remember, I have been fascinated by the natural world, in all its intricacies. I first learned about climate change by watching nature documentaries with my family, but it wasn't until I became older that I realized the broader scale of the issue. The crisis is expected to cause catastrophic damage to homes and livelihoods, cause up to 25 per cent of species to become extinct, acidify our oceans and melt the polar ice caps.

As young children, we were told to turn the lights off, recycle, walk instead of drive, but as kids in our generation grow older, we are able to see that these small actions do not result in the critical change that is essential for our planet's future. This change is only possible if our governments and economic sector come together to tackle the looming crisis. Our message to world leaders is clear and simple – listen to the scientists and put the future of this planet, the future of our home, above all else.

VANESSA NAKATE, 23

Uganda

Vanessa Nakate (born 1996) is a youth striker from Kampala, Uganda. She has been doing weekly strikes for a long time. When she saw that while the burning of the Amazon and California were in the news all the time, there was no mention of the Congo Rainforest crisis, she knew that someone had to stand up for it. It is the second-largest rainforest in the world, but many companies are destroying it and the lives of its animals and Indigenous communities. Vanessa began to strike every day, and the campaign #SaveCongoRainforest reached more global attention. Vanessa striked for over 150 days, and then had to stop for a while because she was so busy. However, she still tries to strike for it when she can.

Previously, a similar campaign, #SaveCongoForest_Flora_Fauna, had been started in the Democratic Republic of Congo (DRC) by Remy Zahiga, another activist, and his friends, but Vanessa made the cause more widely known and also initiated the daily strike for it. The two campaigns are now effectively one, with Vanessa and *Remy being credited as co-founders.*

Vanessa also founded the Rise Up Movement, and Youth for Future Africa, two groups aiming to empower African climate activists and promoting their voices. Additionally, she organizes and manages sustainable projects such as installing solar panels and institutional stoves in local schools.

As if she hadn't already done enough incredible work, in April 2020 Vanessa launched a podcast and YouTube channel dedicated to amplifying the voices of other climate activists from around the world. And then, a couple of months later, she started 1 Million Activist Stories, to share exactly that – stories of youth climate activists from all over the world.

Her country has suffered devastating floods and landslides, giving Vanessa all the more reason to fight. She is staying strong. And she believes that she, and all climate activists, will change things.

Vanessa made it to COP25 in Madrid in November 2019, where she met Greta Thunberg and gave an incredible speech, as well as several interviews.

Vanessa rose to more global recognition at the World Economic Forum 2020, when she was cropped out of a photo with several white activists, including Greta. She was devastated, but this event is not what defines her. She was an activist long before it happened, and she will be for a long time after.

At COP25, she was referred to as 'the Greta Thunberg of the Global South'. But Vanessa is not a Greta, although she is just as inspiring. Because we don't need to be Greta. This isn't just about one person, however awesome. Because, as Vanessa says, 'Every country has a climate activist. Every climate activist has a voice. Every voice has a story to tell. Every story has a solution to give. And every solution has a life to change.'

So Vanessa isn't the Greta of the Global South. She is Vanessa Nakate, climate activist from Uganda, Africa. And she is amazing.

INSTAGRAM: @vanessanakate1 / TWITTER: @vanessa_vash

Hello! My name is Vanessa Nakate, and I am a climate activist from Uganda. I started striking in 2019, by going to the streets with my placard. I also started to strike to save the Congo rainforest, and that is a strike that has gone on for many days now. The Congo rainforest is the largest in Africa, and it is home to various species of animals and plants. Millions of people heavily depend on the forest, and on all of us to protect the forest from any destruction.

Becoming a climate activist was a process for me. In the year 2018, I wanted to do something that could cause a change in the lives of the people in my community. I researched, to understand the problems faced by the people in my community. That climate change is one of those problems was a surprise to me. I had been in school, and all along the teachers had made me, and other students, think that climate change had already happened in the past or that it was coming in the far future, and that we didn't have to worry about it. I decided to read more about climate change, to understand the causes and the impacts. I then realized that some of these consequences were already being experienced in some rural communities. For example, the Mount Elgon area in Uganda has experienced torrential rainfall, leading to flooding and landslides. I have seen the victims cry out to the government for help after losing everything about their life and survival. That is, loss of their houses, loss of lives and loss of their farms, which are the only source of survival. I see the pain as they cry out for help, because of the devastating and frustrating disasters that leave them with completely nothing. Most of the people heavily depend on subsistence farming as a source of survival, and this clearly explains that climate change greatly affects the availability of food, leading to food scarcity. There is also a problem of water scarcity in areas that are hit with intense droughts. I have seen the number of street children increase in Kampala, and I have personally observed that these children come from the same region which has always been known to experience semi-arid conditions. With the increase in global temperatures, these places become uninhabitable for these children and their parents. This shows how much climate change is changing lives and forcing people to leave their homes and travel to the city for paradise, only to find out that they have to beg on the streets to survive.

As a climate activist, I want world leaders to step up their game. They need to stop risking our lives only because of profits that benefit small populations. People are suffering, while others are dying. Now is the time for world leaders to take responsibility and protect our planet. Our home is in jeopardy. Our future is uncertain. We are not sure of our survival. World leaders need to stop thinking that we still have time. We don't have any time left. The people on the front lines of the climate crisis can't live any more, because of their decisions. We lost time the moment these people started dying and suffering. We need to change now. We need action. We are tired of empty words and promises. We want to see real change happening. We want a better world. We want a better future, and it is possible to get it. All we need is for leaders to take charge and clean up the mess they caused. We need to save the future now. The present is already messed up. World leaders need to completely give up the fossil fuel industry. It is possible to transition to more sustainable paths to development. We want to see more sustainable cities, breathe in clean air, and study knowing that the future is certain. Now is the time for world leaders to shift from temporary pleasures to things that will save our lives and secure a cleaner and healthier environment.

The climate crisis terrifies me. It's more like, or even worse than, a horror story. I have dreams and hopes. I have things that I want to achieve in my life, but how can I live normally and work on achieving my dreams with a future that is so uncertain? I am not okay. I am worried. I am scared. I don't want to be a victim of climate change. Every time I find out that more people have died because of climate change, it breaks my heart. We need to put an end to this now. I want my normal life back. I want to feel that my future is secure. Everything is just frustrating.

The thought of waking up, and everything being completely gone, is also very disturbing. This crisis is not good for anyone's mental and physical health.

Every activist in the climate movement is very inspirational. Knowing that you are with so many people, demanding the same thing, is so motivating and encouraging. Sometimes it feels like no one is listening. It feels like I am all alone, but when I remember the other young people striking for a better future, it gives me peace and hope. That is how I keep moving forward. My advice to fellow activists would be that they should not give up, no matter what happens.

Everyone is important and respected in this movement. Every voice matters in this movement. Keep demanding for action. Keep striking. We are together in this and we shall win. When you feel alone, try to reach out to me or any other activist. There are millions of people cheering you on. There are millions of people supporting you. We love you and we are with you. Don't ever feel alone. To anyone who would like to join climate activism, everyone is welcome. Find a local group near you that you can work with. You don't need to start from scratch. We are a global movement. We work together. We support you. We are with you.

YAĞMUR OCAK, 13

Turkey

Yağmur Ocak (born 2006) is a teenage youth striker and climate campaigner from Istanbul, Turkey. She strikes every week, sometimes with friends, like Deniz Çevikus (whose essay appears earlier in this book). She is also one of the co-founders of Fridays For Future Turkey.

For ages, Yağmur wondered about this thing called 'climate change', and would watch lots of videos about things like 'sustainable life hacks' and zero waste. One day, her mum saw her watching one of these, and told her about Greta Thunberg and her school strike. Interested, Yağmur started researching Greta, and discovered her TED talk and her other speech videos. Then she heard about the first global climate strike and was inspired to strike in her school to raise awareness of the issue among her classmates.

Then she striked again a few days later, and after that she was a regular FFF striker and a committed climate activist.

One year after her first strike, which happened to be on International Women's Day, Yağmur reflects on this. The climate crisis is a gender issue because, in many ways, women (and other non-men) will be affected more than cisgender men. This is true for many reasons. Firstly, what with societal collapse, which will occur if we don't act, issues like discrimination and abuse will become much worse. This means that women, and other genders which aren't male, will be discriminated against even more than before. Secondly, it is often true that women and other non-men are less privileged, and more trapped at home or in a small community, meaning that they feel the effects more when that community is hit. Thirdly, women have often been those who are the guardians of the land. And that land is hurting. That land is suffering.

Also, without equality, our systems will just be as destructive as ever. That's why the climate crisis is a gender issue. And a social justice issue.

As a girl, and a child, Yağmur knows that her voice is often seen as unimportant, and she doesn't let this stop her. Because no matter what some people might think, it is. And no matter how much they try to stop it, Yağmur, and millions of others around the world, are changing things. We are mobilizing for global climate justice. And we will not be stopped. Least of all Yağmur Ocak. No matter how toxic, the words of the haters will not stop her. The opinions of the powerful will not silence her. Because her voice is strong enough to defy them all.

INSTAGRAM: @_yagmurocak_ / TWITTER: @ocakyagmur1

My name is Yağmur Ocak and I live in Istanbul. I have always been interested in issues like environmental pollution, recycling and zero plastic. But I came to realize how big an issue the climate crisis really was when I was introduced to Greta Thunberg by my family. They heard about her on the radio, and told me when they came back home. The subject immediately caught my attention. I listened to her speeches, did a lot of research about her, and decided to do a climate strike.

I did my first strike on 8 March 2019, which was also one of the first strikes in Turkey. I gave a lot of information to my friends at school. I then participated in the first global climate strike on 15 March. The attendance at this strike was not very high, but the people I met there, and the interest of the media, was promising. Soon after this, I realized that the interest of the mainstream media was a temporary one. Therefore, we had to continue doing the strikes and make sure to keep the climate crisis a hot topic. Together with my friends, whom I met during the global strike, we formed Fridays For Future Turkey. At the meetings and environmental events we attended, we talked about Fridays For Future, the climate crisis and about Greta. We also met the mayor of Istanbul, wrote a few articles for the few interested media, and took part in radio shows.

Personally, I did my best to explain the climate crisis to those around me at school, and in my social life. I shared photos and articles through social media. I warned everyone I could about over-usage of plastic, and the importance of recycling. I tried doing strikes at school, sometimes on my own, sometimes with friends, sometimes during school days, sometimes during my holidays. I also made sure to participate in all global climate strikes.

I used my banners during strikes to give general information about the climate crisis, as well as current affairs such as wildfires in the Amazon and Australia, Black Friday consumption madness,

COP25, and so on. Sometimes I changed tactics and wrote 'Ask me whatever you want' on my banners to create curiosity, and answered people's questions.

It was a great coincidence that 8 March 2020 was not only my first year anniversary of being a climate activist, but also International Women's Day. That Friday, I had performed my thirty-third climate strike. Considering the fact that women are more affected by the climate crisis than men, it is no wonder that women play a vital role in climate activism, and in fighting the system which is based solely on fossil fuels.

Where there is no gender equality, there is no climate justice. Justice is a very important concept in the climate crisis. The effects of the climate crisis are not only felt by people based on their gender, society or social groups – as a matter of fact, underdeveloped countries, poorer societies and women are the most affected. These groups unfortunately happen to be those who least contribute to the climate crisis.

Additionally, it is very unfair that humans see themselves as the sole rulers of the world, and think they have the right to damage the flora, fauna and nature overall. This world belongs to all species. However, due to our actions, we are creating the end of the world that houses millions of different species. The recent outbreak of coronavirus has proved that we, as humans, aren't as strong as we think we are. We were also helpless against some of the natural disasters we had last year, like wildfires and flash floods, all of which happened as a direct result of climate change. Imagine that these natural disasters start happening all around the world, just like the spread of coronavirus. Just like viruses, the climate crisis knows no boundaries. We will not be protected from this crisis just because our borders are closed, or our country is under quarantine. The leaders must learn from all these crises and start acting together. A good way to do this would be to start listening to scientists, and to eliminate carbon emissions altogether.

In order to go to a high school, I must pass a very important exam soon. I may fail if I don't study very hard. But because of the climate crisis that adults are accountable for, I have difficulty concentrating. I am worried about my future. Instead of studying, I am inclined to think about the climate crisis. I can't help thinking that if my future is in so much danger, what is the point of studying? I do enjoy doing climate strikes. I work on these issues, read about them or write about them, but I am overstressed due to my exams. Adults and leaders have no right to put me under so much stress and anxiety. Instead, they should concentrate on finding solutions to the climate crisis. I am sad, angry and worried.

Me and all my climate activist friends, from all around the world, must continue striking, educate others about the climate crisis, and encourage people to take action, until we see concrete results. To those who want to act together with us: read and learn first and start educating others. Never think that you alone cannot make a difference. Always remember that Greta started her strikes all by herself. Leaders and decision makers: in the future, you will not be sitting on those chairs. It will be us who will be making decisions. Listen to scientists, and act responsibly. We are the future, and we will never give up!

ZOZO AND KIWI, 8 AND 11

Germany

Zozo and Kiwi (born 2011 and 2008 respectively), known in this book only by their nicknames, are sister and brother. They come from Germany. And they're known as the 'Kletterkinder' or 'climbing kids'.

Even before Fridays For Future started, this amazing pair tried to raise awareness of the climate crisis, for example 'by dressing up as polar bears and going swimming with a melting ice floe', at the climate summit COP23, which was held in Bonn.

In January 2019, they climbed a tree in front of the meeting place of the German Coal Commission to display a banner as a protest against coal, which is a huge contributor to greenhouse gas emissions and therefore to global heating. Having been climbers all their lives, it seemed an obvious way for them to demonstrate about the issue.

In June of the same year, they took their courageous exploits still further, abseiling down a bridge during an international FFF demonstration in Aachen. It was at that time that someone started using the hashtag #Kletterkinder, so they decided to use that for their Twitter account, and the name stuck.

In September, during the massive global climate strikes, they climbed the front of the Brandenburg Gate in Berlin. And at COP25, in Madrid, they hung a banner from a bridge during the FFF protest, and climbed a mast at the entrance to the conference building.

It was footage from Madrid that inspired me to ask them to be included. An inspiring video of an eight-year-old girl (Zozo) climbing a lamppost! I instantly wanted to know who she was, and someone must have told the 'Kletterkinder' because they soon got in contact with me, and I was amazed to find out that there were, in fact, two amazing young 'climbing kids'!

Their message is that the climate crisis is far more dangerous than their climbing. So, adults, if you're truly concerned for their safety, stop fussing about heights. The climate crisis is what you should be worried about. So, follow Zozo and Kiwi's advice, and act now.

TWITTER: @Kletterkinder

Isn't this Dangerous – Kids Abseiling down a Bridge? No, the Climate Crisis Is Dangerous

When we abseil down a bridge or climb a lamppost to hang up a banner, people ask us: 'But isn't this totally dangerous?' Of course, this question is not completely unexpected. We actually want people to ask themselves where the dangers are.

Climbing, however, is not dangerous. The climate crisis is.

OKAY, we have to say this here: dear adults, please don't try this at home. Of course you have to know what you are doing. You have to know about redundancy and abseil devices, about knots and breaking loads. We are lucky to have rescue climbers as parents. That's why we cannot even remember when we started climbing. The technique we use is called 'industrial climbing'. You always use two separate systems (e.g. ropes). If one breaks or you mess up with one, the other one will still hold you. That's why climbing can be done safely.

What we don't understand is why everyone else feels safe, even though we are heading with full speed into the biggest crisis of humankind.

The climate crisis is already here. People are starving because of droughts, people are dying in floods, and the forests are burning everywhere on the planet. Even in our safe and protected home in Germany, we can see it. We live in a small village and go climbing in the trees pretty often (who would have thought it?). But last summer, a whole spruce forest just died in the drought. On one day, all the trees lost their needles. A couple of days later, the bark of whole trees fell off. We talked to the local forest warden. He said that even the beech trees are starting to get sick. For millennia, our region was covered in primeval beech forest. Beeches are what define our forests. And now the forest warden doesn't know what to plant any more.

We cannot understand why people still feel safe with this. Do we really need more forests burning or more people dying before we realize that we have to change things big time?

Another question that people ask us is this: 'How old are you? Eight and eleven? Did anyone tell you to do this? Did your parents tell you to do this?' It's strange. For some people it seems impossible that children can think for themselves. Of course, we ask for help where we need it. And of course, we bring our rescue-climbing parents with us when we climb. That is what professionals do: always have a rescue scenario.

We guess that there are three possible reasons to ask that question. Maybe they actually haven't understood the urgency of the climate crisis. OKAY, we can help there, it's not really hard: there is a certain amount of CO_2 that the world may produce and still have a 67 per cent chance of staying below 1.5°C. This amount has been calculated by the IPCC. If we don't make really, really big changes in how we live, this budget will be used up in eight years, or at least this was true at the time of writing, in December 2019. Most scientific studies since 2015 show that we'll reach many tipping points between 1.5°C and 2°C. One example of a tipping point is the permafrost soil. If that thaws, huge amounts of methane will be released. With these tipping points, we might quickly end up in a 3, 4 or 5°C world. And in that world, the question will be if there are millions of people dying or billions. That's terrible, but not hard to understand.

But maybe people take their anger out on children who speak their minds, because it would be too hard for them to accept the consequences. Maybe they just want to go on driving their SUVs?

Or maybe they are like the German minister for environment, Svenja Schulze. We had the chance to ask her about the CO_2 budget for 1.5°C. Of course, she didn't answer. She explained, in many words, how wonderful the German and European plans to reduce CO_2 are. She certainly must know the IPCC numbers, right? If she does, she must know that her plans are by far not enough for the 2°C goal, let alone the 1.5°C goal, right? So she must know that her plans will cause hunger, death and war in the future. So anyone who reminds them of what they should already know must be very annoying. Or maybe some people want to distract the attention from the fact that they are the ones telling others what to do?

Did you know that Fridays For Future activists were thrown out of COP25 because they went up to the stage and chanted together with Indigenous people? And did you know that the climate summits, hosted by the United Nations, are sponsored by fossil fuel companies? Sponsored like a soccer match. And guess who was not thrown out of COP. Guess who could still tell the politicians what to do.

That's why we are the Climbing Kids. That's why we won't stop speaking up. That's why we won't stop annoying them.

Our parents call it non-violent direct action (NVDA). What it means is: don't let them stroke your hair, and tell you how sweet you are, and don't let them lie to you with a smile on their face about what they pretend to do against the climate crisis. But be better than them. Know the science and confront them with it. Be angry, but don't be rude. And if you have to break the rules a little bit for a greater good, remember: you are responsible for it, responsible that no one is harmed.

Interestingly, though, we have found out that there is no law that explicitly forbids children from climbing up lampposts or abseiling down bridges.

ACKNOWLEDGEMENTS

There are so many people to thank in my life that I could write a whole different book on it. But since I can't, I'll just mention some of them here and hope that they know that this is not the extent of the thank-yous I could give.

To start with, everyone who helped with the book. Firstly, to everyone who wrote one of the pieces which this is all about, thank you for your thoughts and words, for your time, for your support, and for your advocacy. Keep fighting.

A huge, huge thank you to Susie Nicklin, for all your tireless work for this, for your belief in me and this book, for getting it going and for never underestimating me. It's always been my dream to be published, and thanks to you, it came true.

To Honor, Juliet, Saliann, and everyone else at Indigo who made *Tomorrow Is Too Late* what it is.

To Andrew Robinson, for your great advice as regards the admin processes for this book.

To Helen, Maude, Natalia, Anna, Daniel, Patsy, Lucy and Kyle, once again. There are literally no words to describe everything you all mean to me. You are the best friends anyone ever had, no debate, and you are *all awesome*. Again, no arguments. I love you all so much. Thanks for all the laughs, support, care and amazingness, as regards this book and everything else. I don't know where I'd be without you. Y'all are the brightest lights in my life. Never ever think you don't matter. You, each of you, deserve the freaking world. Also a special shout out to Kyle for helping me when I got writer's block and for being there when I got stressed about this. You're a star.

To my incredible dad, Mark Maddrell, for being the 'parental permission' I needed for various points of this book, for encouraging me, for all your advice and support of and belief in me, and for everything else. And to my amazing mum, Rebecca Eedle, for your support, for listening to me read from this book in its earlier stages, and for constantly doing all those everyday things that no one really notices, but that without which I could not do anything that I do. Love you both.

To all my other activist friends, too many to mention, but a special shout-out to Elsa, you are amazing! And so many others – if I've called you my friend, then this is you. Thank you so much for your endless love and support. And a specific thank you to Vanessa and Remy for your incredible work for the Congo rainforest. A special shout-out to Annabelle, too, for all your endless support of me.

To my non-activist friends, past and present, especially Hannah, Maia, Ellie, Bella H and Arlo.

Thank you to everyone in the LGBTQ+ Twitter community who has helped, inspired and supported me. I am 100 per cent sure I wouldn't be half as open and proud as I am today without you.

To my younger brothers, Joe and Johnny. I know we fight and argue but you can be pretty cool sometimes, when you try, and good or bad, I can't imagine life without you... peace and quiet, what is that? And to my older brother Ol for all your help, advice and support, and – this goes for all three brothers – all the times we've talked about *Star Wars*. Thanks also to my cousin Echo for your endless and unconditional support of me; Granny, for eternally supporting my writing; and to all my other cousins, on both sides of the family, and to my aunts and uncles, and also to Grandma and Grand. I miss you both.

To Mrs Smith, one of my oldest and dearest friends. I miss you so much, thank you for being like a third grandmother to me.

To many of those who have supported or encouraged me. Again, a few I have to shout out to: Zhiru, Martin, Ann, Alan and Mrs Finch, who helped make my life at both my schools more bearable. To most of my neighbours, in both places I've lived.

To all the amazing animals in my life, especially our hamsters, Ferris and Ares.

Thanks to Lauren Sharkey, who put together *Resisters*, which helped ignite the idea for *this* book in my mind.

And to so many of those who, every day, dedicate themselves to fighting for what's right. Especially to those who have given up so much, who fight in countries where it is dangerous to do so. You are incredible. Although this doesn't do you justice, I am so grateful for so many who stand up for what's right, who helped with this book, and/or who make my life liveable every day.

Transforming a manuscript into the book you hold in your hands is a group project.

Grace would like to thank everyone who helped to publish *Tomorrow Is Too Late*.

THE INDIGO PRESS TEAM

Susie Nicklin
Alex Spears
Phoebe Barker
Honor Scott

JACKET DESIGN

Michael Salu
Andy Soameson

PUBLICITY

Claire Maxwell

FOREIGN RIGHTS

The Marsh Agency

EDITORIAL PRODUCTION

Tetragon
Sarah Terry
Alex Middleton